A Teacup in a Storm

A Teacup in a Storm

An Explorer's Guide to Life

MICK CONEFREY

Collins

This edition first published in 2005 by
Collins, an imprint of
HarperCollins*Publishers*
77–85 Fulham Palace Road
Hammersmith
London W6 8JB

Collins is a registered trademark of
HarperCollins*Publishers* Ltd

everything clicks at **www.collins.co.uk**

1

Copyright © Mick Conefrey, 2005
Images Copyright © Conefrey Films, 2005

Illustrations by Adam Burton except
p. 97, 115 – Edward Whymper, *Scrambles Amongst the Alps*
p. 75, 127, 128 – Elisha Kent Kane, *Arctic Explorations*
p. 126, 142, 175 – Richard Burton, *The Lake Regions of Central Africa*
p. 108 – Ernest Shackleton, *South*

A catalogue record for this book is available from the British Library

ISBN 0-00-720398-5

Typeset by Rowland Phototypesetting Ltd,
Bury St Edmunds, Suffolk
Printed and bound by Clays, St Ives Ltd

To Phyllis, Charlotte and Frank Angelo

Contents

Acknowledgements

I'd like to thank all the people who helped me to get this book off the ground. I'm very grateful to Philip Parker from HarperCollins for taking the project on, and to David Palmer for taking over the reins with such skill and enthusiasm. I'd also like to say a big thank you to Leah at Gillon Aitken, and to my indomitable agent Anthony Sheil. Tim Jordan, Hugh Thomson, John McAvoy, Amanda Faber, Jochem Hemmleb, Harry Taylor, Peter Hillary and Stephen Venables all helped me out at various points with books and information. I'd like to say a special thank you to Adam Burton for agreeing to illustrate the book: his work, as ever, is exemplary. Finally, I owe a huge debt of gratitude to my darling wife Stella who has supported me, put up with me, bought sweets for me and in so many, many ways helped me throughout this project.

Mick Conefrey
March 2005

Introduction

All expeditions are unique, but many follow an archetypal pattern. Someone sets a goal, recruits a team and then prepares to leave. They set off and then spend time in the field, overcoming obstacles and crises. Finally they stop, having succeeded or failed in reaching their goal. Then they turn back; some return to fame, some to indifference and some never return at all.

In this book, *A Teacup in a Storm*, I've tried to find a different way to look at the history of exploration, to present it through a kind of 'how-to manual' in order to examine its paradigms and principles. I've included plenty of advice, based usually on historical incidents, but this book is not intended as a survival manual or a practical guide to the ins and outs of modern-day exploration. What interests me is not so much the detail of individual journeys but rather the general patterns.

Chapter one, 'Getting Started', looks at the first stage on any expedition: organising a team, choosing equipment and, most importantly, raising funds. Exploration is an expensive business and there have been few explorers who haven't had to put a lot of time and effort into finding sponsorship. Chapter two, 'Getting Going', looks at the so-called 'objective hazards' of expedition life: climate, weather, terrain, wildlife and all those constraints that the physical environment throws up. Chapter three, 'Getting Along', focuses on the human side of expedition life. Facing down a charging lion requires obvious courage and self-belief but putting down a mutiny or assuaging hostile natives is just as demanding. In chapter four, 'Getting There', the book changes gear, focussing in depth on six expeditions and trying to discover why some failed and others succeeded. The final chapter, 'Getting Back', looks at that double-edged sword – fame – and the vital issue of keeping control of your own story.

In choosing examples, I've concentrated on the 'heroic age' of exploration: the nineteenth and early twentieth centuries. In this period, exploration increasingly came to be seen as a 'pure' activity, justified by abstract notions of 'Science', 'Discovery' and the needs of the 'Human Spirit', rather than by claiming some commercial purpose for travel. This was also a period when explorers and mountaineers invariably felt it necessary to record their adventures and misadventures in print, so it is much easier to find first hand accounts of the great expeditions.

There is a bias in the book towards colder regions; this is partly to do with personal experience and inclination, but it is also the case that, in general, mountaineers and polar explorers tended to operate in larger groups, so it is much easier to discuss issues such as leadership and team-work by looking at expeditions to the Arctic and the Himalayas, than by looking at the exploration of Africa for example. Wilfred Thesiger, Heinrich Barth, Freya Stark, Richard Burton: they were all fascinating characters, but most of their travel was done in the company of local guides rather than other explorers.

Exploration is a very specialised activity but it is also one that frequently resonates with everyday life. Few of us will ever have to deal with that charging lion or elephant or, for that matter, tuck in to penguin stew but when it comes to planning, crisis management, dealing with strangers, and many other human issues, there are obvious lessons from the history of exploration, which apply to both the slopes of Everest and to the wider world.

A Teacup in a Storm

Benefactor

Destiny

Debts

Risk

Frien

First

Inspiration

Funds

Glory

Trav

Ambition

Gear

Bank

Fame

Com

Team

~ 1 ~

Getting Started

Am going to cross the Pacific on a wooden raft to support a theory that the South Sea Islands were peopled from Peru. Will you come?

> A MESSAGE FROM THOR HEYERDAHL TO KNUT HEGGELAND, TORSTEIN RABY AND HERIK HESSELBERG INVITING THEM TO TAKE PART IN THE KON-TIKI EXPEDITION. THEY ALL SAID YES

Men wanted for hazardous journey. Small wages, bitter cold, long months of complete darkness, constant journey, safe return doubtful. Honour and recognition in case of success.

> ADVERTISEMENT PLACED BY ERNEST SHACKLETON IN A LONDON NEWSPAPER, AUGUST 1914. IT IS SAID THAT 5,000 PEOPLE RESPONDED TO IT

Small minds have only room for bread and butter.

> ROALD AMUNDSEN, ON BEING ASKED WHY HE WENT TO THE SOUTH POLE

What makes an explorer?

Explorers come in all shapes and sizes, literally and metaphorically. Henry Morton Stanley was small and stocky, Fridtjof Nansen was tall and thin, Fanny Bullock Workman was round and tubby. The American explorer Robert Peary advised in his book *The Secrets of Polar Travel* that the ideal Arctic explorer should weigh 2–2 1/2 pounds (0.9–1.125 kilograms) for every inch of height and warned that tall men were a liability – they ate too much, needed too much clothing and took up to much space in a tent. He also added that whenever possible he had selected blondes for his expeditions, ignoring the fact that on all of his Arctic trips he had been accompanied by an Afro–American manservant Matthew Henson. Obviously Henson wasn't in on the secret.

It is equally hard to pin down the archetypal explorer. Some have been tyrants, others team players; some were solitary, others gregarious. Reading through the biographies of nineteenth- and twentieth-century explorers, it is hard to find many common threads. On the contrary it is striking just how diverse their backgrounds were. They began their careers as engineers, missionaries, sailors, soldiers, journalists and beekeepers . . . the list goes on. Quite a lot came from the military for the obvious reason that the armed forces were keen on exploration. More intriguingly, a number of explorers came from troubled family backgrounds and several of them lost their fathers at an early age. Almost all of the polar explorers were men; female explorers seemed to be more interested in Africa and the Middle East.

Many were motivated by the desire for fame. At the age of 17, Edward Whymper, the Victorian mountaineer, confided to his diary that he was hoping to become 'the great person of my day'; on the other side of the Atlantic, at the age of 29, a desperate Robert Peary admitted in a letter to his mother that he simply '*must* have fame'. A lot of explorers professed in later life to have felt a sense of destiny, a sense that they were cut out to

The moon landing	NASA	1969	$25.4 billion
The first ascent of Everest	British Expedition	1953	£20,000
The first ascent of McKinley	Hudson Stuck	1913	$1,000
The first crossing of Greenland	Fridtjof Nansen	1888	£275
The first crossing of Australia	Burke and Wills	1860	£57,840

do important things. Others admitted that they 'liked the life'; when the Scottish explorer John McDouall Stuart returned from his attempt to cross Australia, he had become so used to sleeping outdoors that he scorned his bed and took to sleeping in the garden.

Some explorers never hung up their rucksacks; others found that they could use their skills to carve out successful careers in completely different fields. Quite a few former explorers finished life as politicians. Their fame got them a foot in the door but their skills as communicators and organisers turned them into successes. Many of today's mountaineers and explorers have made a lucrative career as motivational speakers, spending more time speaking to businessmen than lecturing at the Royal Geographical Society. It is easy to be cynical about this, but to do so is to miss the point that, essentially, all expeditions are projects that have to be managed. Teams have to be selected, equipped and motivated. Goals have to be set and crises have to be overcome. The physical risks may be greater and the rewards may be less tangible, but it all comes down to knowing what you want and giving yourself the best chance of getting it.

So what makes a good explorer? Adaptability, ambition, stamina, self-belief, doggedness, curiosity, optimism, authority, hardiness ... but before all these come into play, you have to be good at raising money.

Raising Funds

I shall tell you what you will do. Draw a thousand pounds now; and then when you have gone through that, draw another thousand; and when that is spent, draw another thousand; and when you have finished that, draw another thousand, and so on; but FIND LIVINGSTONE.

J. GORDON BENNET'S INSTRUCTIONS
TO H. M. STANLEY

When J. Gordon Bennet, the famous proprietor of the *New York Herald*, sent Henry Morton Stanley off in search of Dr Livingstone, money was no object. There are few explorers who have had it so easy. Funding is, and always was, a perpetual problem. Getting over this first hurdle requires ingenuity, perseverance and luck. It is one of those times where your people skills come to the fore: how do you persuade benefactors to part with large sums of cash to enable you to realise your dream?

PAYING YOUR OWN WAY

The Duke of Abruzzi paid his own way, and that of a large retinue of supporters, to the Arctic, the Mountains of the Moon and the slopes of K2. He was the grandson of the King of Italy and money was rarely an issue. The British explorer Samuel Baker had a large private fortune which he used to fund his expeditions to Africa. He even paid for his own wife, Florence, after bidding for her at a Turkish slave auction. Their personal wealth gave Abruzzi and Baker a lot of freedom but few twentieth-century explorers were so lucky; at one point or another most of them had to tap an external source for money.

WHAT YOUR COUNTRY CAN DO FOR YOU

The Royal Navy funded many of the most glorious episodes in the history of British exploration. Cook, Vancouver and Flinders are just three of the naval officers whose voyages of discovery helped to re-draw the map of the world. But the navy also funded some of the most inglorious episodes, the polar expeditions of Franklin and Nares being well-known examples. The problem was that naval patronage invariably came with strings attached. Patrons had to be indulged, officers had to be taken on who weren't suitable for the job, equipment had to be used because 'that's how they did it' in the navy. On a smaller scale though, the military is often worth approaching by independent explorers for equipment and sundry favours. Thor Heyerdahl was able to persuade the British and US Military to donate equipment and rations to the *Kon-Tiki* expedition by offering to road-test them in the mid-Pacific.

RICH BENEFACTORS

The best type of benefactor offers you a large sum of money and then retires into the shadows surrounded by a warm, but discreet, glow. When Fridtjof Nansen conceived of an expedition to cross Greenland, initially he was confident of a grant from the Norwegian government. However, there were so many letters to the press proclaiming that he would certainly die on this foolhardy mission that the government said no. Out of the shadows came Augustin Gamél, a Danish philanthropist who had funded previous Arctic expeditions. He covered the whole cost of the expedition and didn't ask for anything in return.

The reward for some benefactors was to see their names appearing on the map. The Boothia Peninsula in the Arctic was named after Felix Booth, a London gin magnate, who underwrote Sir John Ross's second expedition to the Arctic. Booth also donated several bottles of gin to the expedition, one of which was used to christen its ship, the *Victory*.

THE BUSINESS OF EXPLORATION

There has always been a commercial side to exploration. The kings of Spain and Portugal funded the voyages of Christopher Columbus and Vasco da Gama because they hoped to reap their rewards in gold. Exploration and empire went hand in hand. When the Victorian mountaineer Edward Whymper made his epic trip through Ecuador, none of the locals could believe that he had come for climbing and scientific research. They were convinced that he was really on a treasure hunt. On his return to Britain, Whymper was pestered by another species of get-rich-quick entrepreneur, the British businessman. He received numerous letters from tyro tycoons who wanted to hear about all the mineral deposits that, so they assumed, he must have found 'in them hills'. Years later Whymper was employed by the Canadian–Pacific Railway to walk the length of their track in order to publicise the Rockies as a tourist destination. By then he was keener on alcohol than exploration but the railway men still saw his presence as an asset. On a more sinister note, in the 1880s Henry Morton Stanley was employed by the King of Belgium to explore the Congo basin in order give an air of geographical legitimacy to King Leopold's covert empire building.

THE ADVANTAGES OF A COMMITTEE

When in 1965 the British explorer Wally Herbert began raising funds for a trans–Arctic expedition, initially he didn't do too well. He pored over *Who's Who* and sent out hundreds of letters but he received very little in return. Then he got a committee behind him. It was made up of some of the most venerable figures in British exploration, and was quickly set to work. Within three days of its first meeting, Herbert had an office, a secretary and a car. The committee was also happy to share its wisdom and offer its advice. Most of this was gladly taken, but a few months into the expedition Herbert locked horns with his elders and betters over what to do with a sick member of his team. The committee wanted the man evacuated; Herbert wanted to keep him on the ice. Their contretemps was leaked, prompting a spate of articles in the British press. Herbert accused the committee of not knowing 'what the bloody hell they were on about'; they replied that he was clearly suffering from 'winteritis'. It all blew over quickly but the brief spat showed that committees do have their downsides.

TALK TO YOUR FRIENDLY
BANK MANAGER

Modern-day adventurer, Catharine Hartley, the first British woman to walk to both the North and South Poles, spent months trying to raise sponsorship for her Antarctic trip from every business source she could think of, but to no avail. At the point when she was about to pull out, her bank manager rang up and offered an interest-free loan of £10,000. He had heard about her plight after coming back from a Himalayan walking holiday and was feeling generous. It was a stroke of luck, but it wouldn't have happened if Catharine had not spent so many months giving interviews and publicising her plans.

TAP YOUR FRIENDS

When Hiram Bingham decided to put together the expedition to Peru that ultimately would result in the discovery of Machu Picchu, he simply talked his friends and colleagues at Yale into coming along and paying for themselves. He was a famously charismatic and charming man, and ended his days as a politician. The discovery of this sacred city of the Mayans set him up for life and ensured him easy sponsorship for subsequent expeditions.

ROBERT E. PEARY: THE GREATEST FUNDRAISER
IN THE HISTORY OF EXPLORATION?
* * *

There are many people who do not believe that Robert E. Peary got all the way to the North Pole in 1909. That may or may not be so, but there is no doubt that he was the greatest fundraiser in the history of twentieth-century exploration and that his approach is well worth studying. He deserves this award not because of the amount of money that he raised but because of the many and varied ways that he found to do it. For 28 years Peary was a serving officer in the US Navy, but for 16 of those years he was on leave, either on, or preparing for, an Arctic expedition. During all this period, bar six months, he was on full pay. When he came back from the Pole in 1909 he was promoted to Rear Admiral, much to the annoyance of many of his colleagues. Peary's naval salary covered his domestic expenses but he still had to raise thousands of dollars to fund his expeditions. Early on in his career he spent a lot of time on the lecture circuit; in one year he made 165 appearances in 103 days, making $20,000 in the process. In order to fuel interest in his expeditions he was always bringing things back from the Arctic and putting them on show. At various times he returned with huskies, Eskimo artefacts and the second largest meteorite in the world. He took consignments of Eskimo bones to the American Museum of Natural History in New York and even brought back seven live Eskimo 'specimens', some of whose bones also ended up at the museum after they succumbed to TB.

As his fame grew he amassed around him a group of rich patrons, the self-styled Peary Arctic Club. They pulled favours with President McKinley and, through their contacts, helped Peary to become a friend of President Roosevelt. In return, Peary literally put his supporters on the map: Morris Jessup and Herbert Bridgman both had Capes named after them and George Crocker gave his name to an island and a range of mountains (which subsequently turned out not to exist). In 1904, the Peary Arctic Club raised $130,000, then a massive sum, to commission a purpose-built ship for Peary's final two attempts to reach the North Pole. In 1909, as he headed north across the ice, Peary filled his diary with endless speculation on all the commercial endorsements he might attract if he reached his goal and indeed when he returned there were many:

- Winchester rifles
- Shredded Wheat breakfast cereal
- Koh-i-nor pencils
- Rubberset shaving brushes
- Brunswick thermal underwear
- Thermos flasks
- Phillips watches

Equipment

1 Henry rifle, 1 Winchester, 1 double-barrel shotgun, 1 elephant rifle, 1 Starr's breach loader, 1 Jockelyn rifle, 24 muskets, 6 pistols, 1 battle-axe, 2 swords, 2 daggers, 1 boar spear, 2 American axes, 24 hatchets, 24 butchers' knives.

H. M. Stanley's armoury for his expedition to find Livingstone

60 bottles of vin ordinaire, 10 bottles of St George, 15 bottles of St Jean, 3 bottles of brandy, 1 bottle of cassis, 6 bottles of lemonade and 2 bottles of champagne.

Albert Smith's drinks list for his ascent of Mont Blanc in 1851

25,000 cigarettes, 284 boxes of matches, 120 batteries, 100 candles, 100 bulbs, 100 tool kits, 30 torches, 16 lbs of tobacco, 6 hurricane lamps, 3 spring balances, 2 altimeters, 2 pair of binoculars, 1 hairdressing set.

Miscellaneous items taken on British expedition to Kanchenjunga in 1955

The African explorer David Livingstone once wrote that the key to successful travel was to carry as little 'impedimenta' as possible. This philosophy suited him perfectly as he rarely had any money to buy impedimenta, even if he had the urge. For other explorers and mountaineers, choosing the correct equipment and clothing has been a vital first stage to their expedition. Successful teams are invariably well equipped: the British Everest expedition of 1953 was scrupulously kitted out and clothed and this undoubtedly contributed to its success and brought other mountaineers calling for advice on how to tackle the remaining Himalayan giants.

The best equipment though is not necessarily the most expensive or the most technically complicated. When Hudson Stuck climbed Alaska's Mount McKinley with a small party of locals in 1913, much of their gear was home made. Stuck had attempted to import bona fide mountaineering equipment from New York but it either failed to arrive or, when it did, was found to be entirely inappropriate. With a characteristically Alaskan love of denigrating anything that came from the 'Lower 48'(the Alaskan nickname for the rest of the United States), Stuck declared that the imported ice axes 'resembled children's toys' and the hobnail boots were 'useless'. He got a local blacksmith to hammer out crampons and ice axes and used locally made moccasins insulated with five pairs of socks for the high-altitude climbing.

When it comes to equipment, although all expeditions are different, there are certain basic principles that apply across the board:

KEEP IT FLEXIBLE

When Fridtjof Nansen and his party attempted to make the first crossing of Greenland, he carefully planned his equipment so that he could get as much out of everything as possible. The ground sheet from his tent doubled as a sledge-sail and ended its life as the base for an improvised boat.

KEEP IT LIGHT

The world's lightest crampons weigh just 14 ounces (390 grams); the lightest snow shovel is 21 ounces (594 grams) and the lightest ice-axe, the Cassin Ghost, weighs in at just seven ounces (200 grams).

PAY ATTENTION TO DETAIL

David Hempleman-Adams refers to the humble zip as one of the most important elements of an Arctic explorer's outer clothing. It is essential that pockets can be opened without gloves having to be taken off. Zips need to be conscientiously lubricated with graphite because conventional lubricants freeze in cold temperatures. The difference between a zip that works and one that doesn't is frost-bite.

DON'T TAKE TOO MUCH

When Burke and Wills set out from Melbourne in 1860 aiming to make the first crossing of Australia, everyone knew they had taken too much stuff with them. They had eight tonnes of food and supplies, six tonnes of firewood and 440 pounds (200 kilograms) of medical supplies. They even carried a bathtub and an oak table with a matching set of chairs. Unsurprisingly nearly all of this was abandoned by the time they reached their half-way point.

DON'T FOOL YOURSELF INTO THINKING THAT HAVING THE BEST EQUIPMENT IS A SUBSTITUTE FOR HAVING THE BEST PLAN

Salomon Andrée's attempt to reach the North Pole in a hydrogen balloon in 1897 was one of the best-equipped expeditions in polar history. He and his two companions had the world's first primus stove, ultra-modern aluminium cutlery, a Hasselblad camera and an onboard darkroom. Their balloon was rigged with an ingenious system of drag ropes that were supposed to make it steerable and it even had a small cooker, which could be ignited remotely. Everything looked great but there were fundamental problems: in the cold Arctic air the balloon leaked hydrogen and it was prone to accumulating frost on top of its canopy. After just three days, the balloon sank down, depositing Andrée and his two colleagues onto the ice. Thirty-three years later their bear-chewed bodies were discovered on White Island, a miserable strip of land less than 200 miles (322 kilometres) from where they started and, at 80°N, 600 nautical miles (1111 kilometres) from the North Pole.

THE EQUIPMENT THEY COULD HAVE DONE WITHOUT
✳ ✳ ✳

Abruzzi's Bed When, in 1897, the well-heeled Duke of Abruzzi set off for Mt St Elias in the remote wilds of Alaska, his camp equipment included four iron beds. A man who might one day have become the King of Italy was not going to be allowed to sleep on the floor with the guides and the porters. In fact the Duke discovered that with all the cold air circulating under the foot-high beds, it was actually warmer to sleep on the ground. Halfway through the expedition the Royal sleeping apparatus was abandoned.

Shackleton's Car When Ernest Shackleton left for the Antarctic in 1907

on the *Nimrod* expedition, he carried on board his ship an Arrol–Johnston car. It had been donated for free and according to the company's press release, there was a fair chance that it would 'sprint to the Pole'. This was the first time that a car had been taken to the Antarctic, but in the event it took a lot of time and energy to keep the Arrol–Johnston going, and for little reward. Out on the ice, the car tended to overheat, leaving Shackleton and his men waiting around in the freezing cold for it to cool down.

The Sourdoughs Pole In 1910 a group of Alaskan miners, 'Sourdoughs', carried a 14-ft (4.2-m) flag pole up to the North Summit of Mount McKinley. They hoped that it would be visible in Fairbanks 300 miles away but because of atmospheric conditions no one could see the summit, never mind the flag pole, and the Sourdoughs were widely disbelieved. Two years later the flag pole was spotted by Hudson Stuck's team, thus vindicating their claim; however, Stuck also discovered that McKinley had two summits and that the other one was 850 feet (250 metres) higher. The Sourdoughs had been the first men to climb McKinley but they had erected their pole on the wrong summit.

Freuchen's Dinghy Danish explorer Peter Freuchen once punctured his rubber dinghy whilst out fishing in the Arctic. Somehow he managed to get a fishhook stuck into its rubber skin; when the dinghy eventually capsized he was lucky to get away with his life. On another occasion Freuchen almost died

of exposure when the zip on his tent froze fast so that he wasn't able to get inside. If only he'd used a kayak and an igloo . . .

Hunt's Mortar The British 1953 Everest expedition was equipped with a 2-inch mortar for use as an avalanche gun. The theory was that if Hunt and his

men encountered any slopes which were dangerously burdened with snow, they would let fly with the mortar in order to create a controlled avalanche. The mortar was never used for this purpose but on the way back from the mountain, they did fire off the 12 mortar bombs in an impromptu firework show.

THE EQUIPMENT THEY COULD HAVE DONE WITH
✳ ✳ ✳

Wiessener's Crampons In 1939 the German climber Fritz Wiessener turned back from what seemed liked an easy climb to the summit of K2 after his climbing partner, Pasang Lama, dropped their crampons. If they had

had spares, Wiessener was sure that they would have made history as the first men to climb a 26,000-foot (8,000-metre) peak.

Burton's Boat In 1856 Richard Burton and John Hanning Speke left Zanzibar for a journey into the interior of East Africa. They were on a mission to find the fabled source of the Nile and hoped to investigate a group of lakes which had been reported by Arab Traders to lie hundreds of miles inland. Their equipment included an iron boat which they intended to use to explore the lake regions but because they had so much difficulty in hiring porters, the boat was abandoned. When they arrived at Lake Tanganyika they spent many frustrating weeks trying to borrow or hire a boat and were never able to investigate the lake properly.

Courthauld's Snow Shovel In the spring of 1931, August Courthauld, the British polar explorer, became trapped in a dome tent *under* the snows of Spitsbergen because he inadvertently left his snow shovel outside the tent door. He survived on his own for six months before a relief party arrived to dig him out.

Izzard's tent In 1953 Ralph Izzard, an intrepid journalist working for the *Daily Mail*, was sent to Nepal to cover the British Everest expedition. The organisers had signed an exclusive contract with *The Times* and were wary of any other press men, so Izzard's trip was strictly unauthorised. He bought most of his equipment at the bazzars in Calcutta and, when fully equipped, set off on the long march from Katmandhu to Everest base camp. On the first night out Izzard's porters inadvertently destroyed his tent, ripping a large hole in the roof. For the next five weeks he slept in a 'pup tent' made by an American toy company, which was so small he 'projected' at both ends. Izzard, though, was used to discomfort: he made most of the journey in a pair of plimsolls.

Putting a Team together

Getting the team dynamics right is crucial to the success of any expedition. Though a lot of African explorers chose to go it alone, preferring to use native help rather than surround themselves with European assistants, until relatively recently mountaineering and polar exploration invariably required large teams. When it comes to putting a party together, the first thing to say is that temperament is as important as ability. For a team to work properly there needs to be a clear sense of a shared goal and a willingness on the part of all members to muck in for the common good. Of course, this is much easier to write on the page than it is to practise in the field. The stress of taking part in an expedition can stretch team spirit to the limits.

Selecting a team is not always easy. Public advertisements often prompt thousands of applications but it is a haphazard way to choose people. The best teams seem to come from people who know and like each other before joining up, though this isn't often possible. The problem with interviewing unknown candidates is that personality traits often don't appear immediately. In 1935 the British mountaineer Eric Shipton was preparing to go to Everest when he spent a weekend with a friend, Lawrence Wager, who was putting together an expedition to Greenland. Wager told Shipton that he had come to the conclusion that it was always important to include someone in a team that everyone else would dislike; it would give them a focus for their complaints and help them bond. Wager was considering taking Michael Spender to Greenland as his surveyor, and expedition bogeyman. Michael Spender was a notoriously difficult character and Wager was convinced that no one would like him. Shortly afterwards, much to his consternation, Eric Shipton was informed by the Royal Geographical Society that Michael Spender was in fact being attached to *his* Everest party, in order to make a photogrammetric survey of the North Face of Everest. It didn't bode well, but the RGS was sponsoring the Everest expedition and, anyway, Shipton was mischievously interested in seeing what would transpire. In the event, he and Spender got on famously; they became close friends and went on several other expeditions together.

Going it alone was always the preferred option of Wilfred Thesiger. The great explorer of Africa shunned European companions on most of his travels. In 1933 he undertook his first expedition to Abyssinia with David Haig-Thomas, a former school friend. Haig-Thomas was easygoing and cheerful, but he suffered a succession of illnesses. When he announced that he couldn't complete the trip, Thesiger simply carried on going. For the next 40-odd years he stuck to the same pattern, roaming Africa and the Himalayas in the company of local guides. The solitude suited him.

Final Preparations

TRAINING

The mother of Alexander Gordon Laing, the first British explorer to reach Timbuktu, liked to tell journalists that her son used to prepare for expeditions by sleeping on a hard floor. He also taught himself to write with his left hand and, in case of real emergency, by holding a pen between the toes of his right foot. This kind of dedication was unusual and, in general, modern-day adventurers tend to spend much more time in the gym than their predecessors ever did. In the past, exploration was a slow, stately affair. It took George Mallory two months to get from Darjeeling in India to Everest base camp in 1922; much of the latter part of the journey was done on foot. Today's mountaineers can fly into Tibet and then take a jeep to the bottom of the mountain. It has therefore become much more important to be physically prepared before you arrive; there is no longer the option of getting fit on the walk in. The polar expeditions of Scott, Peary and Amundsen involved months and sometimes years in the field. In contrast, when David Hempleman-Adams and Rune Gjeldnes walked to the North Pole in 1998, it took them 55 days to get from their base on Ellesmere Island to 90° north. David Hempleman-Adams' training regime had involved dragging a bundle of car tyres around the Swindon countryside, spending evenings in an industrial deep freeze and regular 10-mile (16-kilometre) hikes carrying a rucksack filled with 80 pounds' (36 kilograms) worth of weights. Other aspects of his preparations had been more pleasant though: in order to put on two stone he spent regular nights down his local pub downing pints of lager and bags of crisps.

THE SEND-OFF

There have been many great last-night parties, which have resulted in a bedraggled and slightly hung-over team arriving at the docks or the airport on the following morning, with a few fond, but dim, memories of the night before. Official send-offs, though, are a different matter. The biggest and most flamboyant events often seem to precede the most disastrous expeditions. When Burke and Wills left Melbourne 15,000 people, over a tenth of the city's population, turned up to see them off. The great and the good struggled with the great unwashed to glimpse the expedition's camels and wagonloads of equipment; there were bands, speeches and much drunken revelry and then the expedition was off, never to return.

AND INTO THE FIELD ...

Once the money has been found, the team chosen, the equipment sourced and the final preparations completed, it's time to go into the field. In the next stage explorers will have to deal with the physical hazards of altitude, climate and terrain, they will have to fend off wild beasts, meet and greet natives, learn how to put up with solitude and pain and decide amongst themselves who does the cooking.

Now the real fun starts.

HOW TO DEAL WITH AN
ANACONDA ATTACK
***

Apocryphal extract from a US Peace Corps manual

1. If you are attacked by an anaconda, do not run. The snake is faster than you are.
2. Lie flat on the ground. Put your arms tight against your sides, your legs tight against one another.
3. Tuck your chin in.
4. The snake will come and begin to nudge and climb over your body.
5. Do not panic.
6. After the snake has examined you, it will begin to swallow you from the feet. Permit the snake to swallow your feet and ankles. Do not panic.
7. The snake will now begin to suck your legs into its body. You must lie perfectly still. This will take a long time.
8. When the snake has reached your knees, slowly and with as little movement as possible reach down, take your knife and very gently slide it into the snake's mouth between the edge of its mouth and your legs. Then suddenly rip upwards, severing the snake's head.
9. Be sure you have your knife.
10. Be sure your knife is sharp.

(This set of instructions was widely circulated on the Internet in the late 90s and even made its way into acclaimed screenwriter William Goldman's book, He Lied – More Adventures in the Screen Trade. *Anacondas are very dangerous snakes and have been known to eat entire wild pigs but there are very few records of anacondas actually killing humans. Like all snakes, they can eat their prey whole and, because they can unhinge their jaws, they can swallow prey much larger than the diameter of their mouths. However they usually swallow the victim's head first so that the legs don't get stuck on the way down.)*

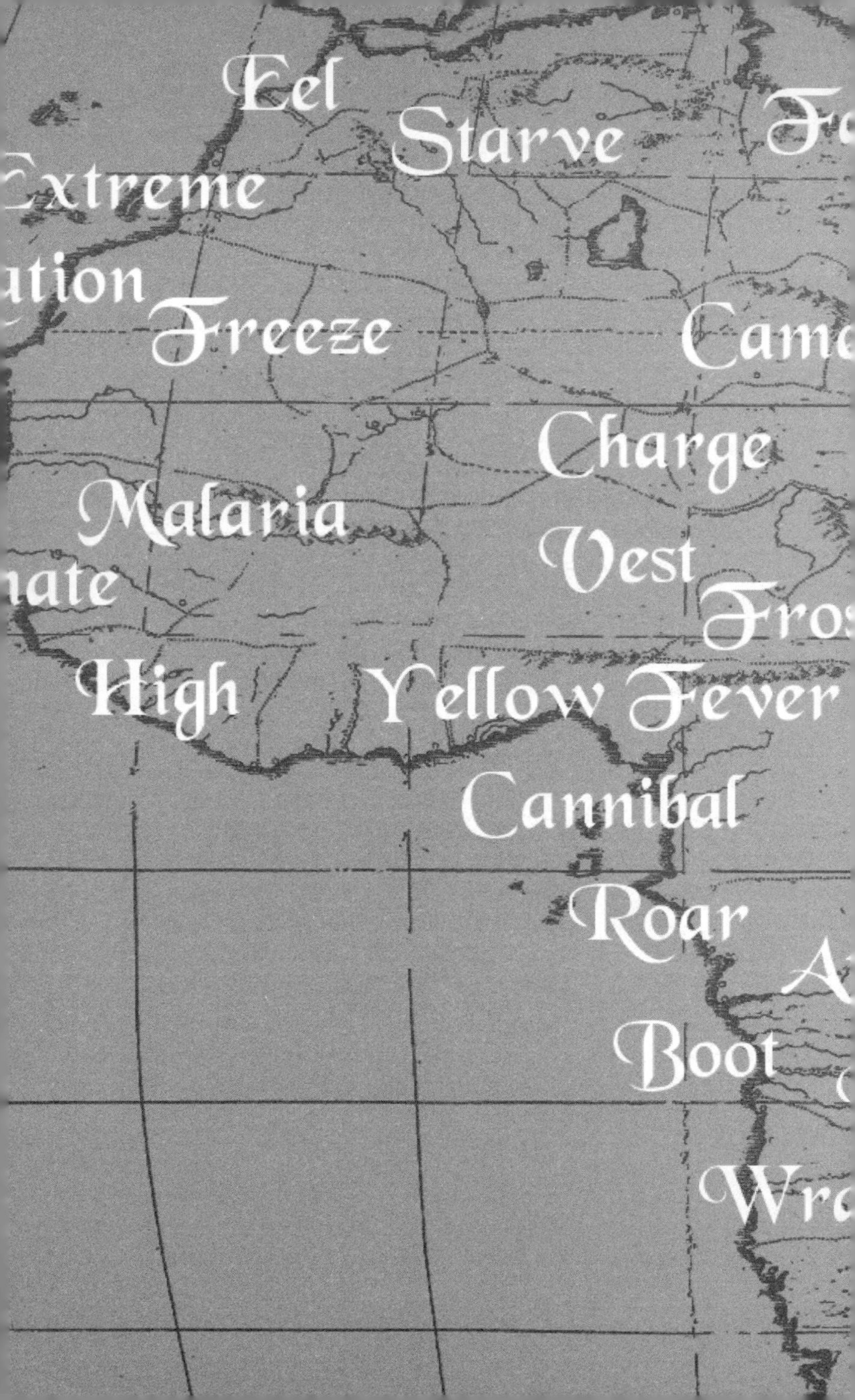

Eel

Starve

F...

Extreme

...tion

Freeze

Cam...

Charge

Malaria

Vest

...ate

Fros...

High

Yellow Fever

Cannibal

Roar

A...

Boot

...

Wr...

Getting Going

What a sad difference there is from what I am now and what my party was when we left North Adelaide. My right hand nearly useless to me by the accident from the horse; total blindness after sunset and nearly blind during the day; my limbs so weak and painful that I am obliged to be carried about; my body reduced to that of a living skeleton, and my strength that of infantine weakness – a sad, sad wreck of former days. Wind variable.

JOHN MCDOUALL STUART, ON THE WAY BACK FROM HIS EPIC CROSSING
OF AUSTRALIA IN 1862

It seems as though we are in some other world, and yet the things that concern us are most trivial such as split lips and big appetites.

At one moment our thoughts are on the grandeur of the scene, the next on what we would have to eat if we were let loose in a good restaurant.

ERNEST SHACKLETON, DURING HIS ATTEMPT
ON THE SOUTH POLE IN 1908

No one knows the value of water till he is deprived of it … I have drunk water swarming with insects, thick with mud and putrid with rhinoceros urine and buffaloes dung, and no stinted draughts of it either.

DAVID LIVINGSTONE

LEAVING HOME ...

Your leaving day might not seem like a particularly risky moment, but it is important to stay alert at all times. When in the spring of 1947 a tug arrived to tow the *Kon-Tiki* out of harbour at the beginning of its epic voyage across the Pacific, Thor Heyerdahl, the expedition leader, was the only man on board. The other five members of his team were still ashore, having a final drink at the bar and doing some last-minute shopping. What followed was an hour of high farce: the tug almost crashed into the raft and sank it, the tow rope broke, the other men were informed that the *Kon-Tiki* had sailed, and at one point the raft filled up with beautiful women brought along by the Peruvian Navy. Eventually, the *Kon-Tiki* did manage to float out into the open sea. 'If it starts this badly,' quipped one of Heyerdahl's companions, 'then the rest of the expedition is bound to be a doddle...'

Few expeditions are a doddle. Exploration *is* full of risks and dangers. Foremost amongst those are the 'objective hazards': these are the dangers of terrain and environment. For desert travellers there is the heat and the difficulty of finding water; for mountaineers there is the peril of high altitude; for polar explorers there is the problem of operating in the cold. With the right approach and the right equipment these environmental hazards can be eased though they can never be totally overcome.

High

THE SEVEN SUMMITS

The first person to climb the highest mountain in every continent was an American businessman, Dick Bass. His best-selling book inspired many

Everest	29,029 feet (8,708 metres)	Asia
Aconcagua	22,840 feet (6,852 metres)	South America
Denali	20,320 feet (6,096 metres)	North America
Kilimanjaro	19,339 feet (5,801 metres)	Africa
Elbrus	18,481 feet (5,544 metres)	Europe
Vinson Massif	16,067 feet (4,820 metres)	Antarctica
Carstenz Pyramid	16,023 feet (4,807 metres)	Australasia/Oceania
Kosciuszko	7,310 feet (2,193 metres)	Australia

imitators, though many serious climbers were quick to point out that some of the mountains were very easy. There is some dispute over the seventh highest mountain: is it Kosciuszko in Australia, or the much more imposing Carstenz Pyramid in Indonesia? Just to be on the safe side, some climb both.

High altitude is not good for you. In Tibet and Peru there are people who reside permanently above 10,000 feet (3,000 metres), but most of us do not handle the high life too well. The problem is lack of oxygen: as the altitude increases, the air pressure decreases, which means that it becomes harder and harder to breathe. The air pressure at the summit of Everest is three times lower than at sea level so your lungs have to work much, much harder to take in enough oxygen for your body to function properly. If you can't get enough oxygen, hypoxia ensues: oxygen starvation.

The generic term for the physiological problems caused by high altitude is 'acute mountain sickness' or AMS for short. It is not just a problem at extreme altitude. At 4,500 feet (1,350 metres), just over the height of Britain's tallest mountain, Ben Nevis, the first signs of AMS can start to show: trekkers and climbers may find themselves slightly out of breath and some may get headaches and feel nauseous. At 10,000 feet (3,000 metres), the height of quite a tall peak in the Alps, these symptoms can be much more pronounced. At 20,000 feet (6,000 metres), the height of a very tall mountain in Alaska, things get much more serious: AMS can develop into cerebral and pulmonary oedema, two conditions which are potentially fatal. At 26,000 feet (7,800 metres), a mountaineer is well into the so-called 'Death Zone'; breathing is very, very difficult and according to some experts, prolonged exposure makes permanent brain damage almost inevitable. All of which means that anyone who manages to survive a trip to the summit of K2 or Everest is very lucky to get back down.

Of course, many hundreds of people have done so and there are many mountaineers and explorers who never had any severe problems with altitude. But it is all very unpredictable. In 1954, a year after making his famous first ascent of Everest, Edmund Hillary was stretchered off Makalu after collapsing at 22,000 feet (6,600 metres), a full 7,000 feet (2,100 metres) below the summit of Everest. He was suffering from a cocktail of pneumonia, malaria, broken ribs and altitude sickness. In 1991 Hillary had an attack of cerebral oedema at a Sherpa village at just 12,700 feet (3,810 metres) and while today he still returns to Nepal on charitable work he finds it difficult to spend more than a few hours at any high altitude. To add to the unpredictability of things, there are no tests that you can do in advance to indicate how you will perform at altitude. In general older climbers fare slightly better than younger ones. Low blood pressure is an asset and the fitter you are, the better, but that's about as far as it goes. Still, there are precautions that you can take which make your visit to the Death Zone but a temporary one.

TAKE TIME

The tried and tested way to deal with altitude is to take everything slowly. Acclimatisation is the process of adapting to altitude, and in general it is best not rushed. The walk in to a Himalayan peak may seem tedious, but it is a vital first stage on any expedition. Imagine a group of climbers heading for Everest from the Nepalese side. They fly into Lukla airport, at 9,200 feet (2,760 metres), and then up towards base camp, 8,000 feet (2,400 metres) higher. Every day they make a series of short marches, resting frequently and rarely spending more than five or six hours on the trail. Gradually their breathing becomes faster and deeper, their heart rate increases and so does their red blood cell count. At base camp they will probably rest for a few days to further enhance acclimatisation. If everything has gone well they should now be ready to take on the mountain.

After that there are basically two different approaches. Traditionally, high mountains were climbed by 'siege tactics': over a period of weeks, the climbers would set up a series of camps higher and higher up the mountain, gradually moving up and acclimatising to each new altitude before making an attempt on the summit. The old adage was 'climb high, sleep low': after setting up camp 2, you sleep at camp 1, then after setting up camp 3, you sleep at camp 2, until you get to the summit. More recently, top climbers have started to use what they call 'Alpine' tactics: instead of a slow build-up, they arrive at base camp and then climb a few of the lower peaks nearby. Then, when they feel ready and the weather is good, they make a dash for the summit sometimes without any intervening camps on the way.

TAKE OXYGEN

The simplest way to compensate for the lack of oxygen at high altitude is to take your own. It has always been a contentious issue though. In the 1920s when the first British mountaineers attempted to climb Everest, there was a running debate over the ethics of using supplementary oxygen. Some felt that without taking it the world's really high mountains would never be climbed, but others argued that using any 'artificial aids' was unsporting. The treasurer of the 1922 Everest organising committee famously remarked that climbers would be 'rotters' if they couldn't reach 25,000 feet (7,500 metres) under their own steam. Apart from this ethical problem, the oxygen systems used in the 1920s were awkward to use and prone to malfunction. The set taken by George Mallory on his final climb in 1924 weighed about 32 pounds (14.5 kilograms), a huge burden at high altitude. The cylinders were so prone to leakage that when the expedition arrived at base camp, they found that out of the 90 they had brought with them, 15 were totally empty and 24 others were well on the way.

It will never be known if the failure of an oxygen set contributed to the deaths of Mallory and Irvine that year, but there's no doubt that the early equipment could be a liability. By the 1950s, oxygen sets were much lighter and much more reliable, and nobody argued about the ethics of using them, or not vociferously at least. Today's state-of-the-art Kevlar and steel cylinders weigh less than 7 pounds (3 kilograms) and the slopes of Everest and the other Himalayan giants are littered with empties. In recent years, there has been a big effort to get climbers to clean up their act and there have been several expeditions to Everest with the stated purpose of retrieving all the old oxygen cylinders and abandoned equipment.

TAKE DRUGS

There are several drugs that mitigate the effects of altitude, but they all have side effects.

Coca

All over the South American Andes, the local people chew coca leaves and drink coca tea. As well as increasing their powers of endurance on a minimal diet, it helps them to cope with high altitude because it increases the rate of respiration. It is estimated that 75 per cent of the population of the high Andes uses coca regularly.

Nicotine

I remember meeting an ancient climber in the basement of the Alpine Club in London. As another venerable colleague puffed away on a pipe, he delicately balanced a tower of ash on the end of his cigarette. When I asked him if smoking got in the way of mountaineering, he rattled back, 'Don't all climbers smoke?' Once upon a time it was thought by some that cigarettes were good for your throat and some climbers even believed that they aided acclimatisation. Today few doctors would agree. Apart from irritating the lungs and reducing the rate of respiration, cigarettes tend to inhibit circulation, making frost-bite much more likely.

Diamox

Diamox, or acetazolamide, is used to treat a wide variety of conditions from epilepsy to glaucoma; coincidentally it is often carried on mountaineering expeditions to aid acclimatisation. Diamox increases the rate of respiration and thereby makes a climber breathe in more oxygen. Its disadvantages are a crop of potential side effects and the fact that because it acts a diuretic, anyone on Diamox usually has to make more trips outside the tent in the cold of the night if they don't have an unusually large pee-bottle.

Dexamethasone

Dexamethasone, or 'Dex' as they call it in Hollywood mountaineering films, is a steroid which is used to treat a variety of illnesses including arthritis and certain forms of cancer. It is effective in reducing inflammation and is sometimes used on mountains in cases of cerebral oedema. In the 1998 film K2, hysterical guides and their clients fight over syringes of Dexamethasone, but it is a powerful drug and should really be taken only under qualified supervision.

Sick of the Altitude

Acute Mountain Sickness (AMS): The best way to deal with any type of mountain sickness is to descend to a lower altitude. This is usually effective in treating mild cases of AMS and if the period at low altitude is long enough, climbers can frequently go back up again. If AMS is not treated it can become very serious: vision is impaired, muscles weaken and pain sensors don't work properly. There is also a significant mental deterioration: victims think slowly, are prone to memory loss and have severely delayed reactions. Some may become depressed; others become euphoric. In the final stage of AMS, unconsciousness is followed by death.

AMS isn't the only danger of high altitude, though. The severe reduction in air pressure can also affect the body on a very basic cellular level, causing fluid to leak from cells around the brain and lungs. The generic term for this is 'oedema'.

Pulmonary Oedema: This occurs when the lungs fill up with fluid that has leaked out of the network of capillaries around the lung walls. It is quite common but is very dangerous.

Cause: Poor acclimatisation and rapid ascent.

Symptoms: Breathlessness, a gurgling sound during breathing and coughing, lips and edges of the ears turning blue, fast respiration, high pulse.

Treatment: Diamox may help deal with the symptoms, but descent is by far the best treatment for anyone suffering from pulmonary oedema. Some expeditions carry special hyperbaric chambers, commonly called 'Gamow Bags' which, when inflated with air, simulate the effect of a drop in altitude, but these can only offer temporary relief.

Cerebral Oedema: This is less common than pulmonary oedema but it is even more dangerous. It is a complicated condition in which the brain swells because of oxygen starvation.

Symptoms: Headaches, retinal bleeding, difficulty in walking, fast pulse, irritability.

Treatment: Dexamethasone and a hyperbaric chamber will relieve the symptoms, but the only cure is to descend.

Prevention: Good acclimatisation is the best prophylactic for cerebral oedema as it is for all the altitude-related illnesses. As already stated, with mild cases of altitude sickness it is often possible to go down and rest at a lower altitude and then come back up, but this is definitely not advisable with pulmonary and cerebral oedema. Some doctors do not recommend taking Diamox or Dexamethasone prophylactically because of their side effects, but others take a more pragmatic view and accept that both drugs can be used if there simply isn't time to acclimatise properly.

TEAR UP THE RULE BOOK:
REINHOLD MESSNER'S ASCENT OF EVEREST

In the summer of 1978, Reinhold Messner and his climbing partner Peter Habeler stunned the scientific and climbing worlds by making the first ascent of Everest without oxygen. Prior to this, many climbers and altitude experts thought that such a feat was impossible and initially there were some people who didn't believe that Habeler and Messner had really done it. But they had, and this climb revolutionised mountaineering.

Habeler and Messner were used to controversy. They had made their names with daring ascents in the Alps in the 1960s and 1970s. Both men disliked using artificial aids for climbing and in 1975 they made a rapid, two man attempt on Gasherbrum I, in Pakistan, without supplementary oxygen. This 'Alpine-style' ascent generated a lot of controversy. Some climbers and medical experts attacked Habeler and Messner for breaking two of the cardinal rules of mountaineering, namely that supplementary oxygen was necessary at high altitude and that climbers had to acclimatise slowly. These sceptics maintained that anyone who climbed high without it was simply being reckless and laying themselves open to permanent brain disease. Habeler and Messner were undeterred. Three years later they decided to take on the ultimate challenge: an oxygen-free ascent of Everest.

After six weeks on the mountain and one failed attempt, they left their camp on the South Col of Everest at 5.30 a.m. on 8 May. The weather was not perfect but they pressed on, climbing first up to the South Summit, then to the Hillary Step, and then finally along the last 166 feet (50 metres) to the summit. It was agony: Messner later described how his lungs felt as if they were going to burst and his mind seemed utterly dead. Habeler went down straight away, but amazingly, Messner stayed on the summit for long enough to take photographs, shoot some cine film and record his thoughts into a tape recorder.

It was an awesome achievement, but initially there were some climbers who simply didn't believe that they were telling the truth. One group of Sherpas was so sceptical that they demanded an inquiry by the Nepalese government's Ministry of Tourism. A few months later, Messner's critics were silenced when one member of a German team and two Sherpas also made it to the summit of Everest, without supplementary oxygen.

Then to cap it all off, two years later Messner made a second oxygen-free ascent of Everest which broke even more rules. Firstly, he did it in the middle of the monsoon, traditionally avoided because of the heavy snows and secondly he did it solo, an amazing feat for the time.

Today there are many top mountaineers who have climbed big mountains without oxygen. This doesn't mean though that the risks of high altitude are no longer taken seriously. There are still many people who die every year in the Himalayas from pulmonary and cerebral oedema, and plain old mountain sickness still ruins the holidays of many trekkers. Messner and Habeler's great achievement was to demonstrate that there are no absolute rules in mountaineering; some might have thought them reckless but once they opened the door, many others followed.

THE FOURTEEN 8,000ERS

Once the Seven Summits became the holy grail of Texan business-men and middle-aged mountaineers, elite climbers dreamt up another much tougher challenge: scaling the fourteen 26,600-feet (8,000-metre) summits. Everest is common to both lists but the fourteen 8,000ers is a much, much tougher proposition. The first man to climb all of them was, you guessed it, Reinhold Messner.

Everest	29,035	Nepal/Tibet
K2	28,250	Pakistan/China
Kanchenjunga	28,169	Nepal/Sikkim
Lhotse	27,940	Nepal/Tibet
Makalu	27,766	Nepal
Cho Oyu	26,906	Nepal/Tibet
Dhaulagiri	26,795	Nepal
Mansalu	26,781	Nepal
Nanga Parbat	26,660	Pakistan
Annapurna	26,545	Nepal
Gasherbrum I	26,470	Pakistan/China
Broad Peak	26,470	Pakistan
Gasherbrum II	26,360	Pakistan
Shishapangma	26,291	China/Tibet

Hot

Francis Galton advised in his Victorian handbook *The Art of Travel* that sucking a lump of quartz or a bullet can allay thirst. He added that travellers had been known to add a little gunpowder to their food to satisfy their craving for salt. Neither of these measures would be recommended today but there's no doubt that for any explorer extremely hot weather presents a serious problem.

Roughly 12 per cent of the world's surface is classified as dry, arid desert; most of this is found between 18° and 28° latitude in both hemispheres. For any traveller there are two fundamental challenges: how to protect yourself from the heat and how to find an adequate supply of

water. In the desert temperatures can reach 57°C (135°F) during the day and fall as low as −9°C (16°F) at night. Deserts are very challenging.

In normal working conditions an average male needs approximately 3,000 calories of food per day but can survive on as little as 500 for several months. When it comes to water, the body is much less flexible. In temperate regions it is possible to live on two cups of water a day, but in a desert between one and three gallons (4.5–13.5 litres) are required. When the temperature increases, the need for water becomes even greater. At an air temperature of 10°C (50°F) it is possible to survive for 14 days without water; at 49°C (120°F) that drops to between three and four days, even for someone carrying a gallon of water.

Sweating is the body's main response to excessive internal and external heat. When sweat evaporates off the skin, it cools everything below it. It is a fairly efficient process but it is also a greedy one. On a hot day, it is possible to sweat two pints (1.1 litres) of liquid in an hour. This must be replenished quickly otherwise your blood will start to thicken, making it harder for the heart to pump it around the body. Anyone who loses more than 5 per cent of his or her body weight, will begin to deteriorate physically and mentally. When this rises to 10 per cent, he or she will go rapidly down hill, becoming deaf, delirious and oblivious to pain. When moisture loss reaches 12 per cent, death frequently ensues. At the same time, sweating drains the body of electrolytic salts, which are necessary for muscles to work properly. If these are not replaced, salt deprivation causes cramps and, in the worst cases, death.

It is possible to acclimatise, to an extent, to hot environments; within a fortnight the average person's capacity to sweat can triple without any serious effects.

The fundamental principle of desert travel is to avoid any unnecessary loss of moisture. In the accounts of Europeans who set out to explore the deserts of North Africa, you often find references to the frustration of travelling with local guides who are invariably accused of being too slow or malingering. Usually these complaints occur at the outset and then, as the journey continues, the outsiders come to realise the good sense of travelling at a languid pace and making sure that when the opportunities present themselves, men and camels are allowed to drink their fill. The desert is a harsh teacher and the Bedu and other tribes have learnt the lessons of how to survive over many thousands of years.

TIPS FOR DESERT TRAVEL
★ ★ ★

- Travel slowly and avoid any unnecessary exertion. It will only make you sweat more.
- When travelling through a desert it is best to move during the coolest period of the day: dawn and dusk or at night.
- Dampening your clothes will keep you cool but it is important to wipe off any salt which condenses on them; it is highly corrosive and can damage boots and clothing.
- It does not usually pay to dig for water unless you are sure that you will find plentiful supplies – usually the effort required outweighs the benefits.
- To minimise water loss keep your mouth closed and breathe through your nose. Avoid talking and any difficult work which will cause you to breathe through your mouth.
- In summer the ground temperature can be up to 30 per cent higher than the surrounding air, so it's better to rest on a stool or a branch than to sit on the ground.
- It is better to wear more clothing than less – fabric absorbs sweat and prolongs its cooling action.

THE TRAVELS AND TRAVAILS OF
JOHN MCDOUALL STUART

One of the greatest explorers of Australia was a Scottish immigrant, John McDouall Stuart. Between 1858 and 1862 he led six gruelling expeditions to the unknown interior; on the final journey he became the first man to lead a party all the way across the continent and return to tell the tale.

Stuart was an unlikely explorer. Small in size and prone to ill health, he had a fondness for whisky, which bordered on alcoholism. Stuart learnt his bush-craft surveying the land north of Adelaide; he was very good at it and was often employed by local farmers to search for new grazing. Stuart's philosophy was to keep his expeditions small and mobile. He would scout ahead whilst the remainder of his team would make a depot at the last good waterhole. Then when he had found the next viable source of water, he would bring the other men up.

On his first major expedition in 1858, Stuart took an Aboriginal tracker but he deserted half way through; after that Stuart preferred to trust his own instincts. Water was a constant preoccupation; occasionally he found himself trapped by flash floods or held back by muddy ground, but most of the time he was desperately searching for water. It was often a dangerous business: at one point one of his assistants was almost buried alive when a well that he was digging in a river bed collapsed on top of him. The Aboriginal tribes that he met on his expeditions were frequently hostile. As if to add to his problems with the heat, they often set fire to the vegetation around his camps.

Stuart set off on his final trans-continental expedition in a desperate hurry. A few days out he discovered that his party's water bags were so leaky they were losing half their volume every day. Eventually they were abandoned. When finally on 24 July 1862 he and his party reached Van Diemen Gulf on the north coast of Australia, it was water that dominated everyone's thoughts:

Thring who rode in advance of me called out 'The Sea!' which took them all by surprise and they were so astonished that he had to repeat the call before they fully understood what was meant. Then immediately they gave three long and hearty cheers ... I dipped my feet and washed my face and hands in the sea, as I promised the late governor Sir Richard McDonnell I would do if I reached it.

How to find water in a Desert

FOLLOW THE BIRDS AND THE BEES ...

The presence of bees is a sure sign that there's fresh water within a few miles. A bee can cover a mile in 12 minutes and will fly in a straight line to a water source 1,083 yards (1,000 metres) away. Grain-eating birds such as

finches and pigeons need water to survive so they are another good indicator of nearby water. They drink at dawn and at dusk, and tend to fly very low and straight when they are seeking water. Flesh-eating birds such as crows and hawks are not such good indicators, because they can survive much longer without water.

... BUT DON'T TRUST DOGS

According to nineteenth-century expedition lore, any water that a dog or a horse can drink is also fit for human consumption. This is not true; dogs have a much more robust digestive system than humans. If in any doubt, it is best to boil any water found in the field. Then to get rid of its flat taste, aerate it by pouring it back and forth between two containers.

LOOK TO THE TREES ...

Palm trees are a sure sign of water; it usually lies within a few metres of their base. There are water reservoirs present in Australian She-oaks (Casuarinas) and in some species of wattle.

... AND THEIR LEAVES

It is possible to extract water from leafy green plants by tying a plastic bag over them as tightly as possible. As the air inside the bag warms the plants release moisture, which condenses on the inside of the bag. It may not taste very good and only small amounts of liquid will be released, but in desperate straits there might be no other choice.

GET UP AT DAWN

Even deserts usually have a heavy dew which lasts for an hour after sunrise. It can be collected by laying out pieces of cloth and then wringing them into a container. An old Bedouin trick is to turn over at dawn any half-buried stones – they are cooler on the underside and will encourage dew to condense.

LEAVE THE CACTI ALONE

Many deserts in North America contain large cacti, but there's a lot of dispute about their value as an emergency water supply. In the Wild West days, the barrel cactus was said to hold a gallon of water inside, just waiting for thirsty cowboys to take a swig once they had chopped the top off. In fact this is not true, although barrel cacti do have reservoirs of liquid stored in their pith; in emergencies this pith can be chewed or squeezed into a container. But be warned: there are some types of barrel cactus which contain toxic chemicals. Chewing their pith will cause vomiting and diarrhoea and make dehydration worse. The flesh of certain cacti causes hallucinations when eaten – fine if you're looking for a 'head trip' but not for a conventional expedition.

AND WHEN YOU FIND WATER ...

Don't drink it too quickly. Rapid re-hydration can cause the electrolytic salts in the blood, lost through sweating, to become even further diluted. In severe cases this can have fatal consequences. It is a good idea to consume small quantities of salt, but salt tablets are often difficult to absorb and cause stomach cramps.

... DON'T ALWAYS EXPECT IT
TO TASTE GOOD

When in 1946 Wilfred Thesiger crossed the Empty Quarter, an enormous, little-explored stretch of desert in Southern Arabia, he well understood that it was going to be a very risky journey. In the summer the temperature can get up to 80°C (175°F) and it contains sand dunes over 500 feet (150 metres) high. His Bedouin guides were very reluctant to go: not only would they have to contend with hostile tribes, but they knew that the journey from beginning to end would be a fight for survival against the relentless heat and lack of water. Thesiger refused any special treatment: he wanted to eat what his guides ate and drink when they drank. That usually meant a few sips of coffee in the morning and one meal in the evening of rough porridge or a few lumps of unleavened bread smeared with butter. They carried their water in leaky goatskins and rationed themselves to about a pint (half a litre) a day sometimes mixed with sour camel's milk. When they did find ancient wells they usually had to be dug out and the water was frequently bitter and brackish. If their camels refused to drink, they forced it down their throats.

A few days into the heart of the Empty Quarter, Thesiger's party encountered some men grazing their camels. The nearest well was many miles away but these Bedouin had found a patch of rich vegetation, produced by a burst of rain several years earlier. Their sole source of food and drink was camels' milk and they were hoping to stay at this spot for many months. Following their strict codes of hospitality, they treated Thesiger and his party well, offering them a goatskin of milk to take away with them.

Thesiger and his guides pressed on, rising at dawn and stopping late at night, working their way through towering dunes of sand. Once, the camel that was carrying their biggest water bags slipped and rolled over but miraculously the bags didn't burst. When one of Thesiger's guides managed to catch a solitary hare, they cooked it and then drew lots for who would have each portion. Finally after about a fortnight, they reached a well which marked the end of their journey and toasted their success:

I had crossed the Empty Quarter. It was fourteen days since we had left the last well at Khaur bin Atarit. To others my journey would have little importance. It would produce nothing except a rather inaccurate map that no one was ever likely to use. It was a personal experience and the reward had been a drink of clean, tasteless water. I was content with that.

WILFRED THESIGER

Sick of the Heat

Dehydration: An occupational hazard for anyone travelling in the desert. Extremely dry air causes the rapid evaporation of sweat and also dehydrates the mucous membranes that line the respiratory system.
Symptoms: Thirst, dry skin and nasal passage, low heart rate and lack of energy, sunken eyes, shallow breathing, dark urine.

Treatment: Water in small amounts; salt in small doses in cases of severe dehydration.

Prevention: Avoid over-exertion in hot, dry weather; wear loose clothing and lots of it.

Heat syncope: A sudden loss of consciousness suffered by people working in hot environments.

Symptoms: Fainting.

Treatment: Move to the shade, take water and electrolyte drinks.

Prevention: Heat syncope often affects people who are not acclimatised to working in hot conditions. Take it slowly . . .

Heatstroke: This is caused by exposure to excessive heat, but it is not a desert disease. It occurs in regions of high humidity such as the tropics, where there is so much moisture in the air that sweat is not able to evaporate off the skin and thereby cool it. Instead it runs off the body and is much less effective.

Symptoms: Exhaustion, fainting, nausea.

Cure: Move to a cool area.

Prevention: Wear loose clothing, take frequent small quantities of water, avoid exertions in the hottest period of the day.

Wet

Thor Heyerdahl, the Norwegian explorer, used to say that when he was a child he was terribly afraid of the sea. He imagined it pulling unwary swimmers down into its depths, never to be seen again. Later in life he realised that the opposite was true: the sea was a friend who tried to keep swimmers afloat through their natural buoyancy; only reluctantly would it allow someone to sink. In truth the sea can be your friend and your enemy: it has fish to feed you and sharks to eat you; it has storms, sand banks and reefs; it is a natural highway which can become an obstacle course.

Most of the great naval explorers were shipwrecked at one stage or another: Christopher Columbus began his career by being shipwrecked off the coast of Portugal and then went on to lose nine ships on his voyages to the Americas. Until an accurate way to measure longitude was discovered in the late nineteenth century, ships would frequently get lost at sea. Vasco da Gama's famous voyage to India took 90 days longer than he expected because of navigational miscalculations. The Arctic and the Antarctic were notoriously dangerous for shipping. Ernest Shackleton's ship the *Endurance* was just one of many ships to be caught in the pack ice.

On the other hand, it is remarkable how even primitive craft such as Thor Heyerdahl's raft, the *Kon-Tiki*, can make long and successful sea voyages. Heyerdahl followed up his epic trip from South America to Polynesia with the *Tigris* expedition in 1977 when he sailed a ship made from reeds from Iraq to Djibouti in an attempt to trace the sea routes of the ancient Sumerians. For Heyerdahl the sea was always a means to bring people together rather than keep them apart.

Many of the hazards of ocean-going exploration are not to do with the sea itself. It is often the physical isolation of sea-faring which causes many problems: how to carry an adequate supply of food and fresh water, how to maintain the good health of everyone on board and how to maintain morale and discipline a long way from home . . . these are but a few of the issues. Even for the Cooks and the Columbuses, the ever-present threat of mutiny was a reminder of the huge strain that long voyages place on expedition leaders.

Sick of the Sea

Scurvy: It could be said that scurvy and malaria are the two diseases most closely associated with exploration; scurvy is well known as a maritime disease but it is a condition that can occur on land as quickly as it does on sea. For many years it was the scourge of the British Royal Navy and though men such as Captain Cook put a lot of energy into finding a way to prevent scurvy, it wasn't until the early twentieth century that scientists really understood its cause. Scurvy blighted British expeditions to the Arctic and Antarctic, and was also a severe problem for the early explorers of inland Australia. John McDouall Stuart had such severe scurvy on his expedition to cross the continent that he couldn't ride a horse on the return journey and had to be carried on a makeshift stretcher slung between two horses. In his journal he described in vivid detail how his teeth and gums ached so much that he couldn't eat and how he frequently vomited blood and mucus.

Cause: Deficiency of vitamins C and, to a lesser extent, B.

Symptoms: Bleeding gums, loose teeth, difficult respiration, swollen limbs, ulcers, darkened skin, anaemia, depression and, as reported by several eighteenth-century mariners, severe homesickness.

Cure: The signs of scurvy usually appear after about three months of vitamin C deprivation and if nothing is done about it the disease is fatal. The cure is simple though: regular intake of vitamin C. In the late eighteenth century the Royal Navy began issuing lemon and lime juice as an anti-scorbutic but though these fruits are rich in vitamin C, the naval habit of boiling the juice down to make it easy to transport usually killed off all the vitamins present.

Prevention: A good diet rich in vitamin C, found in many fresh foods.

Cold

Though the human body finds extremes of heat difficult, it takes longer to acclimatise to severe cold. The body's internal heating system is not as effective a mechanism for raising temperature, as perspiration is for

The coldest temperature recorded, degrees F (C)

Antarctica (Vostok Research Station, 1983)	−129 (−89)
North America (Snag in the Yukon, 1947)	−81.4 (−63)
Asia (Oimekon, Siberia, 1933)	−90 (−68)
Europe (Ust'Schugor, Russia, date unknown)	−67 (−55)
South America (Sarmiento, Argentina, 1907)	−27 (−33)
Africa (Ifrane, Morocco, 1935)	−11 (−24)
Australia (New South Wales, 1994)	−9.4 (−23)

NAUTICAL SLANG IN
COMMON USAGE
* * *

Over a Barrel – Sailors used to be flogged over the barrel of a gun, hence: in a difficult position.

Know the Ropes – There were miles of rope on a sailing ship; a sailor needed a lot of skill to know where they all were, hence: to know how to operate something.

First Rate – Royal Navy ships used to be rated by how many guns they carried. A first-rate ship had 100 guns, a fifth-rate ship had only 20–40, hence: top class.

Pipe Down – The ship's boson's final pipe was the signal for lights out and silence, hence: be quiet.

Leeway – The lee-side of a ship faces away from the wind, and the leeway is the gap between the ship and the shore, hence: room for manoeuvre.

As the Crow Flies – Ships used to carry crows which would be released if a navigator was unsure of his position, they would head straight for the shore and he would time them in order to work out how far away the ship was from it, hence: the shortest distance.

The Bitter End – The bitts are the part of a ship's bow to which the anchor cable is fastened. If all the cable has been played out you reach the bitter end, hence: the final stage.

Slush Fund – Ships' cooks were notorious operators; one common scam was to scrape the fat from the bottom of the barrels which stored the crew's meat and boil it down. The cook would then take the fatty slush that he had gathered and sell it ashore, then share out the profits, hence: a hidden fund.

Above Board – Above deck and therefore visible, hence: clear and straightforward.

Between the Devil and the Deep Blue Sea – The devil seam was a feature in the deck planking on the edge of the ship, so anyone who slipped could find themselves between the devil and the deep blue sea, hence: in a difficult position.

reducing it. Shivering is an automatic response to cold; it generates heat but only after using up a lot of energy.

The main defence against cold is clothing and shelter, and sufficient calorie intake. This may seem obvious, but for explorers in the field keeping the cold at bay has always been a real problem. The lowest temperatures in the world are found in the polar regions. In many ways they are similar to deserts: there is very little precipitation and they are relatively sterile. From the air, Arctic sastrugi look very similar to desert sand.

Clothing

From the eighteenth century onwards, there have been two principal schools of thought about polar clothing. Some explorers were 'adopters': they basically copied the Eskimo style of clothing. Others were 'adapters': they wore a variation of their own traditional cold-weather clothing. Robert Peary, the American explorer who raced to the North Pole at the turn of the twentieth century, was a strong proponent of the first approach.

Polar Eskimos didn't traditionally wear blanket shirts, nor did they line their bearskin trousers with flannel, but basically Peary and his party were dressed like native hunters. Animal-skin clothing is very effective against the cold because the thousands of individual strands of hair which make up a piece of fur are hollow, allowing the air they contain to act as an insulator. Peary took clothing very seriously; he insisted on always getting the best animal skins and kept a group of Eskimo seamstresses on hand to tailor each item to the individual members of his team. Before setting off for the Pole, he sent his team on long hunting trips. This also allowed them to get used to their clothing and sort out any defects before they began the expedition proper.

The disadvantage of skin clothing is that it is heavy and cumbersome and needs a lot of maintenance. Once inside their igloos, Eskimos used to take their furs off in order to dry them and traditionally it was a wife's role to look after her husband's clothing. For Eskimo hunters, the fact that fur parkas were bulky and unwieldy was not a problem: their style of hunting involved long, patient waits in the snow and when they travelled their dogs did a lot of the hard work. Flexibility was important, but the crucial issue was that clothing should be warm. As Peary was also intending to use Eskimo dog teams as his principal mode of transport, skin

ROBERT PEARY'S OUTFIT FOR HIS ATTEMPT ON THE NORTH POLE IN 1909
* * *

1 short-hooded coat of selected deerskin
1 short-hooded coat of sheepskin
1 blanket or flannel shirt
1 pair of short flannel-lined bearskin trousers
1 pair of bearskin or deerskin boots
1 pair of sealskin boots
2 pairs of polar hareskin stockings
1 pair of bearskin mittens
1 or 2 pairs of deerskin or sealskin mittens
3 or 4 pairs of blanket inner mittens
2 or 3 pairs of deerskin inner soles

ERNEST SHACKLETON'S CLOTHING LIST FOR HIS ATTEMPT ON THE SOUTH POLE IN 1908
* * *

Woollen pyjama trousers
Woollen singlet
Woollen guernsey
2 pairs of thick socks
1 pair of finnesko
Burberry overalls and head covering
Balaclava
Woollen mitts
Fur mitts

clothing was entirely appropriate for his expedition to the North Pole.

The British explorers who went to the polar regions in the nineteenth and twentieth centuries had different needs. Traditionally they were advocates of 'man-hauling', i.e. pulling their own sledges. It was much more physically demanding than running alongside a dog team, so they used up much more energy and sweated much more. They needed clothing that was warm, but it also had to be light and flexible; instead of heavy skins, the British wore wool inner clothing and windproof outer garments.

The only fur in Shackleton's outfit was found in the mittens and the footwear. Finnesko were the reindeerskin boots which were traditionally worn by the Lapps of northern Finland. The same basic approach to clothing was used by Robert Falcon Scott on his expedition to the South Pole three years later.

The problem with this type of clothing was that it was fine on the move, but when the man-haulers stopped, they would get very cold, very quickly if they didn't get their tents up straight away. When exercising, the body produces up to ten times more heat than when resting; and even in the polar regions, you will sweat if you are exerting yourself. Unlike modern breathable fabrics like Gore-Tex, old-fashioned windproof clothing tended to absorb sweat rather than disperse it. This gradually made it heavier and heavier and liable to freeze: the British explorer Apsley Cherry Garrard memorably described how after a few days' man-hauling, his windproofs became as stiff as a suit of armour. Shackleton compared his Burberry overalls to chain-mail.

Today's polar explorers have the benefit of a variety of high-tech fabrics that are designed to be worn in extreme climates. Synthetic fleece is woven in such a way that it wicks away sweat and it dries very quickly. Duckdown jackets and trousers are very warm and incredibly light and easy to pack. The most important principle for cold-weather clothing is that of layering garments. Rather than wearing one heavy item, the idea is to wear a series of layers that can be put on and taken off quickly and easily, depending on what activity is being performed. When it comes to boots and footwear, synthetic materials have largely replaced leather and animal skin. Many different variations have been tried out: plastic boots, rubber boots, heated boots, boots that have to be pumped up to provide an air barrier, boots lined with a metal barrier . . . the list goes on.

For mountaineers, good cold-weather clothing is a priority. Even though the polar regions are colder, there is actually a greater incidence of frost-bite amongst mountaineers than polar explorers because high altitude makes it more likely. Mountaineers need clothing that is warm, wind proof, flexible and resistant to sweat. Tenzing Norgay, who climbed Everest with Edmund Hillary in 1953, once put his success down to the fact that he sweated so little when climbing, which meant that he was much less likely to get frost-bite.

A look at clothing worn on Everest over the last hundred years shows how much things have changed:

GEORGE MALLORY'S GEAR FOR HIS
ATTEMPT ON EVEREST IN 1924
* * *

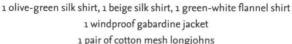

1 fur-lined leather motorcycle helmet

1 long cotton mesh vest

1 brown wool sweater

1 olive-green silk shirt, 1 beige silk shirt, 1 green-white flannel shirt

1 windproof gabardine jacket

1 pair of cotton mesh longjohns

2 pairs of wool leggings

1 pair of windproof gabardine knickers

3 pairs of wool socks

Wool puttees

One pair of felt-insulated nailed leather boots

CLOTHING ISSUED TO THE BRITISH EVEREST
EXPEDITION IN 1953
* * *

Windproof trousers and jacket made from a cotton and nylon mix,
proofed with Mystolen

Leather climbing boots lined with fur and a second layer of leather,
with felt insoles and thin rubber soles

Special high-altitude boots made with a waterproof layer of insulation
and a softer micro-fibre sole

Silk inner gloves, down mittens, windproof outer mitten

2-piece down suit

2 featherweight jerseys and a heavier jumper

String vest

Woollen socks

CLOTHING WORN BY PETER HILLARY IN 2002 ON AN EVEREST EXPEDITION TO COMMEMORATE THE FIFTIETH ANNIVERSARY OF HIS FATHER'S ASCENT
* * *

1 fleece cap, 1 sun hat

1 silk balaclava, 2 polypropylene balaclavas, 1 fleece balaclava

1 neoprene mask, 3 medical masks

6 polypropylene tops, polar and crew neck

3 pairs long johns pants

4 pairs underpants

1 pair of polar fleece pants

1 polar fleece jacket

1 pullover

1 down vest

2 down jackets with Gore-Tex outer shells

1 one-piece down jacket

1 pair of insulated salopettes with a Gore-Tex shell

1 Gore-Tex jacket with additional salopette pants

3 pairs of polypropylene gloves, 1 pair of fingerless gloves

2 pairs Gore-Tex mitts

1 pair of very large gauntlets

3 pairs of polypropylene socks, 4 pairs of thin wool socks, 2 pairs of fleece socks,

8 pairs of heavy wool socks, 1 pair of nylon vapour barrier socks, 1 pair of neoprene

vapour-barrier socks

1 pair of plastic boots, 2 pairs of alveolite inner boots, 1 pair of neoprene over-boots

1 pair large booties

1 pair of gaiters

Just by looking at the size of the differing lists, it is obvious that mountaineering clothing has now become much more specialised. In Mallory's day, climbers wore variations on the jackets and breeches that might have been taken on a shooting trip to the Highlands of Scotland. Each member of the 1924 expedition provided their own personal clothing and they all made idiosyncratic choices. The first man to wear a down jacket on Everest was the Australian climber George Finch who was part of the British team in 1922. It didn't catch on. By the 1950s, down jackets and trousers were considered essential wear on Himalayan expeditions. The vast majority of

Peter Hillary's gear from 2002 was made from man-made fibres such as polyamide and polypropylene and his boots were made of plastic. The latest development in cold weather clothing is the vapour-barrier concept: instead of allowing sweat to permeate through your socks, or sleeping bag, you put an impermeable layer close to the skin. This keeps you warmer because it reduces the amount of evaporation from your skin, leaving your sleeping bag drier and protecting your feet from frost-bite.

Anyone who goes out into the cold with inadequate clothing will soon start to feel uncomfortable, and the longer that person stays out, the worse it becomes. Cold can affect both physical and mental processes, leading to poor decision-making and judgement. It reduces manual dexterity and affects grip strength. Shivering will increase your body temperature but it is inefficient, tiring and makes sleep difficult. When a body gets really cold, it switches on automatic defence mechanisms. Circulation to peripheral regions is reduced, in order to prevent damage to the core organs. This means that your fingers, toes, noses and cheeks all become susceptible to frost-bite.

Sick of the Cold

Hypothermia: A dangerous condition which occurs when the body's core temperature falls to a critical level. A drop of up to 5°C (9°F) is considered mild hypothermia and anything below that is regarded as severe and life threatening.

Symptoms: Numb skin, poor co-ordination, uncontrollable shivering, slurred speech, apathy; in severe cases, unconsciousness.

Treatment: Anyone with mild hypothermia can be helped to warm themselves up, but severe cases need to be evacuated. If this is not possible, they should be gently re-warmed using whatever is available, taking care not to burn their skin and initially paying most attention to the head, neck, groin and armpits.

Prevention: Appropriate clothing, avoidance of excessive perspiration, in extreme weather covering the mouth and nose with an insulating barrier. One golden rule is that, 'No one is dead until they are warm and dead.' Even when someone does not appear to be breathing and has little or no pulse, it may still be possible to revive them if they are warmed properly.

Frost-bite: This occurs when low temperatures and inadequate clothing cause circulation to be reduced to peripheral areas like the fingers and toes. When they reach a low enough temperature the intercellular liquid literally freezes up causing damage to nearby tissues. Like hypothermia, there are several grades of frost-bite: at its mildest there is 'frost-nip'. It is painful when it thaws out but there is usually no long-lasting harm. At its worst, severe frost-bite can require the amputation of a whole limb.

Symptoms: Frost-nipped skin loses sensation and turns pale and waxy, but the underlying flesh is still soft. If unchecked this may develop into deep frost-bite; now the underlying tissue freezes as well. Initially a frost-bitten finger will be very painful but eventually it will become numb and stiff.

Complications: Septicaemia, blood poisoning, gangrene.

Treatment: Frost-nip can be treated by contact with a warm body part, hence all the stories of explorers warming someone else's frozen feet on their bellies. In the past, one of the common treatments for frost-bite was to rub affected areas with snow, the theory being that this would help restore circulation. This doesn't work and risks further tissue damage caused by friction. The best way to deal with deep frost-bite is to immerse the affected limb in hot water between 40 and 42°C (around 105°F), but this isn't always easy in the field. Limbs that have been thawed out and then become re-frosted are at the greatest risk; it is also very dangerous to walk on or use limbs that have been partially thawed out.

Prevention: The key issue is to keep everything covered up and protected by suitable clothing. Mittens and slightly oversized boots are better than tight gloves and perfectly fitting footwear. It is very important to maintain good circulation and to allow fingers and toes to be able to move around. Avoiding washing also reduces the risk of frost-bite, as this helps to keep the skin hydrated.

Famous Frost-Bite Injuries

The history of mountaineering and polar exploration is full of explorers and climbers who have suffered the agonies of frost-bite.

Robert E. Peary made a gruelling journey to a remote hut on Ellesmere Island in the winter of 1898. Arriving there in early January, he discovered

that he was frost-bitten in both feet. Without proper tools or medicine the expedition doctor was forced to amputate seven of Peary's toes, leaving him unable to walk for six weeks. He was undeterred: according to legend, he wrote one of his favourite quotations on the wall next to his bunk: *Inveniam Viam aut Faciam* – I shall find a way or make one. For the rest of his life he had to wear specially built-up shoes which gave him an unusual, penguin-like gait.

Peter Freuchen was a Danish explorer who lived and worked with the Polar Eskimos in the far north of Greenland. In 1923, he was out on a hunting trip when he became separated from his party and was trapped in a blizzard. Freuchen managed to dig a snow hole but when he was rescued, he discovered that one of his feet was severely frost-bitten. Soon gangrene set in and the flesh started coming away around his toes. An Eskimo shaman offered to bite them off in order to keep evil spirits away but Freuchen opted instead to remove them with a hammer. Three years later his whole foot was amputated.

Maurice Herzog was one of two French climbers to make the first ascent of Annapurna in 1950. It was a terrible struggle to get to the summit and an even greater ordeal to get down. By the time Herzog reached the nearest camp, his hands were purple and white and as hard as wood, and his toes were frozen stiff. Two of his team-mates spent the evening whipping Herzog's fingers and toes with rope ends. In the 1950s, this was thought to increase circulation but in fact it just damaged the surrounding tissue. Further down the mountain, the expedition doctor, Jacques Oudot, administered excruciatingly painful injections of Novocaine into Herzog's stomach and then on the trek out, he continued the torture by giving him injections of penicillin directly into his arteries. When they reached Delhi, Oudot discovered that Herzog's feet were infested with maggots, which he was unable to remove, and they remained there until Herzog reached France. There he underwent 12 operations but ultimately none of his fingers or toes was saved.

Beck Weathers was one of the American climbers caught up in the infamous 1996 Everest tragedy. He was given up for dead on two occasions, but managed to survive. He suffered severe frost-bite, losing part of his nose, his right hand and most of his left. Back in America he had ten operations, the longest lasting 16 hours. Amazingly Beck was able to return to work as a pathologist, using voice-control technology and foot pedals to partially compensate for his missing fingers. His experience was life changing in more ways than one; by his own account he became happier and more focused, his wanderlust had gone and his marriage was rescued from the brink.

Humid

Humid regions present very different problems for the explorer. Whereas the deserts, Poles and mountains are generally relatively sterile, the jungles of Africa and South America team with life, and bugs. For nineteenth-century explorers such as Henry Morton Stanley and David Livingstone, diseases such as malaria were taken for granted. Throughout his explorations, Stanley was laid up with frequent bouts of 'fever'; as far as he was concerned, it literally went with the territory. He subscribed to the old-fashioned theory of 'mal aria', or bad air, believing that the disease was some kind of airborne virus which could literally be smelled and tasted in the swamps and jungles of Africa. Whilst exploring the Congo Stanley even went as far as to put up a glass screen on his boat *En Avant* to keep the malarial 'miasma' at bay. Livingstone, a trained doctor, referred to malaria as 'marsh miasmata'. All of his children caught malaria but he never really took it seriously until he was affected himself. Stanley and Livingstone suffered from malaria throughout their lives but other men were not so able to tolerate it: Hugh Clapperton died of malaria whilst searching for the Niger.

Tropical Kit

There have always been clear issues when it comes to cold-weather clothing, but dressing for hot and humid weather has been more eclectic. The British explorer Mary Kingsley went to West Africa in a long black Victorian dress, declaring that she wouldn't wear anything abroad that wouldn't be good enough to wear at home. Fellow Brit Samuel Baker included a tweed suit in his expedition wardrobe and on occasion carried a kilt and a full-dress Scottish highland suit.

The safari suit is a variation on the military clothing introduced in the mid-nineteen century. 'Khaki' comes from a Hindu word for dust and was originally introduced as a clothes dye to help camouflage British troops in India. By the end of the century khaki had been adopted by several other armies. There are many variations of the safari suit but the basic principle is that they should be loose fitting, made of strong enough cotton to resist snags, and have voluminous pockets.

The pith helmet is another product of British India. It is a high-crowned, wide-brimmed hat made from the 'pith', or core, of trees such as cork. This material is strong and durable and can withstand both heavy rain and strong sunlight. Pith helmets were common in India before the British arrived, but colonial soldiers and administrators popularised them and soon they spread around the world.

Sick of the Tropics

Yellow Fever (aka 'Yellow Jack', 'La Vomito'): In the nineteenth century, yellow fever was a major issue for anyone travelling through the tropics; today it is not nearly so widespread, but there are still some countries which demand proof of immunisation against yellow fever before entry. It was first recorded in Africa and subsequently transported by European traders and colonists to South America; there, it ran riot. In the 1880s Ferdinand De Lesseps was defeated in his plan to build the Panama Canal by epidemics of yellow fever. In his first year he lost 500 French engineers brought over to supervise the project. He persevered but finally, after losing one-third of the entire European work force of 20,000, gave up and sold the construction rights to the US.

Cause: Yellow fever is a virus carried by mosquitoes that thrive in swampy conditions.

Symptoms: Early symptoms resemble severe influenza but quickly become much worse. Blood vessels rupture causing splotches of blue and black to form under the skin, then internal organs start to fail. When the liver comes under attack, the skin turns yellow, hence the name. In the final stage, victims vomit blood.

Cure: Yellow fever has quite a high mortality rate, between 40 and 60 per cent. There is no specific treatment apart from rest and plenty of fluids. Anyone with the disease should be away from mosquitoes in case they are bitten and spread the disease.

Prevention: Modern vaccine is safe and effective; other precautions include wearing deet and putting permetharin on clothes.

Malaria: Today we think of malaria as a tropical disease but it wasn't until late into the twentieth century that malaria was eradicated from Europe and the United States. Worldwide, it is estimated that between 1 and 1.5

million people still die every year from malaria, whilst many more are laid low by it.

Transmission: Malaria is caused by microscopic parasites that enter the body through the bite of certain mosquitoes. Once they get into the bloodstream they reproduce rapidly, attaching themselves to red blood cells and subsequently destroying them.

Symptoms: Fever, shivering, joint pain.

Complications: The most dangerous form of the disease is cerebral malaria in which infected red cells obstruct the blood vessels in the brain. Other vital organs can also be damaged, often leading to the death of the patient.

Treatment: Anti-malarial drugs are usually effective but there are some strains of the disease which have now developed a resistance to Chloroquine, one of the most commonly prescribed medicines.

Prevention: There are several drugs that can be taken as a prophylactic; with some of them you have to begin the course before entering the malarial area. The best way to protect yourself is to avoid being bitten. This entails sleeping under nets, wearing protective clothing and insect repellent and avoiding being outside when mosquitoes are most active, i.e. at dawn and at dusk. Public health campaigns in Europe and America have focused on eradicating mosquitoes altogether by spraying swamps with insecticides such as DDT but there are still some areas of the world where malaria is endemic.

Diseases such as yellow fever and malaria make travel in the tropics dangerous but there are many other, macroscopic hazards which also need to be taken into account.

PERCY FAWCETT'S TROPICAL HORRORS

Percy Fawcett was a British explorer who made several journeys to South America at the beginning of the twentieth century. In 1925 he disappeared into the jungle never to be seen again. His journals feature a litany of bugs and beasts that he encountered on the trail:

The puraque, or electric eel: Fawcett noted two species: some were 6 feet (2 metres) long and coloured brown, while others were half the size and coloured yellow. The latter was the more dangerous variety. They killed their victims by giving them repeated electric shocks; natives refused to handle them even when dead. In the eighteenth century, Alexander Von Humboldt, the German explorer scientist, famously conducted experiments on electric eels during his epic trip to South America. If handled correctly he discovered that a single shock was by no means fatal; Humboldt also roped in his assistant to see what happened when two people were in contact with the same eel. He discovered that they both got small shocks.

Apazauca spider: This large spider resembled a tarantula and was reputed to have a fatal bite. Fawcett heard of a rural inn on the Andean Altiplano which had a resident Apazauca in one room. The spider killed traveller after traveller. The natives thought that the inn was haunted but eventually the spider was found, lurking in the rafters.

Candiru, 'vampire fish': This small eel-like fish is famous for its ability to enter the penis and vagina of anyone urinating in a river where candiru are swimming. Usually it parasitises other larger fish, inserting itself under their gill flaps before drinking their blood. With its skin covered in barbed spines, it is very difficult to remove.

Garapatas do chao, 'ticks of the ground': These are tiny white ticks with unbearably irritating bites. When travelling through the Matto Grasso region, Fawcett noted how the garapatas would collect into clusters and hang off tree branches before falling off on to the hapless explorers below. In the evening Fawcett and his companions would pick up to 200 of them off their bodies.

Piranha: This small flesh-eating fish attacks in shoals and can strip the flesh off a body in minutes. Piranhas are particularly attracted to open wounds but Fawcett reported several stories of unlucky natives falling into piranha-infested rivers only to be gobbled up in seconds.

Palo Santo tree: This soft-wood tree grows near river banks; it is a favourite haunt of the Brazilian fire ant. The smallest contact can cause armies of ants to descend on an unsuspecting victim. Fawcett noted that local tribes used the Palo Santo as a weapon of torture, tying a prisoner to the tree for several hours.

Animals

Animals play an important role on many expeditions. They fall into three main categories: pack animals, animals that you can eat, and animals that can eat you . . .

PACK ANIMALS

Horses, ponies and mules

Horses and mules are often used as pack animals. They are strong, reliable and fast, but they can also be feisty, stubborn and difficult to control. The problem with pack animals, or human porters for that matter, is feeding them. If there is little grazing available, the number of animals needed increases because some of them have to carry food for the others to eat. When there is no grazing whatsoever and all the fodder has to be imported, horses and ponies can be a liability. Shackleton and Scott both took ponies to Antarctica for their expeditions to the South Pole; they believed a pony's strength and stamina gave it an advantage over sledge dogs, and that one pony could do the work of several huskies. In theory this was true, but in practice the ponies found it very difficult to work in the snow and ice and couldn't cope with severe weather. They required a huge amount of fodder: Scott took 45 tonnes of it to the Antarctic in 1910, and his horse expert Captain Oates still didn't think that he had taken enough.

Yaks

Wild yaks used to roam all over the Tibetan plateau; today there are very few of them but the smaller domestic yak is common. Yaks are kept for their milk and meat and are also used as pack animals. They graze on herbs and lichens and can be pastured as high as 18,000 feet (5,400 metres). Their thick coats can cope with temperatures as low as −40°C (−40°F) and their large lungs enable them to breathe comfortably at high altitude. Unlike most cattle, yaks are able to sense pasture under the snow and will seek it out themselves. In spite of their large size they are nimble and surefooted and can be trained to follow each other in single file. Long trains of yaks, laden with climbing gear and supplies, have been a common sight on Everest expeditions for the last 80 years.

Dogs

Working huskies are not the cute furry animals that star in Disney movies. They are just one notch away from being wild; they love to fight, they are not reluctant to eat each other and one of their favourite meals is human excrement. These factors aside, they are perfectly suited to pulling sledges across the polar wastes. Until relatively recently dog sledges were the Eskimos' principal means of transport all round the Arctic. The Siberian husky is thought to be the purest breed and is one of the gentlest. Greenland dogs are more aggressive. Alaskan huskies are fast and strong but don't have the same hunting instinct as Greenland dogs, so it is safer

to let them off the leash. Alaskan huskies are mongrels, a mixture of local Eskimo dogs and dogs brought north by settlers. Nineteenth-century polar explorers were divided over the value of dogs: the British were generally not keen on them, but Americans and Scandinavians were quick to copy Eskimo methods. In spite of this, the British were the first to introduce huskies to Antarctica. Robert Falcon Scott used them on his attempt on the Pole in 1902; they were not a success.

Can it be that the dog has not understood the master? Or is it that the master has not understood the dog?

ROALD AMUNDSEN, NORWEGIAN EXPLORER ON WHY
SCOTT DIDN'T LIKE DOGS

When in 1901 the British explorer Robert Falcon Scott headed south on the first major British Antarctic expedition, none of the members of his team had any experience of dogs. Not surprisingly, Scott and the others got on singularly badly with their huskies. On board ship there were several tonnes of dried fish which had been brought as dog food; it was not stored properly and went off on the long voyage down. By the time Scott's ship reached the Antarctic, several of the dogs were sick and refusing to eat their feed. Undaunted, in November 1902 Scott set off for the South Pole with a team of 19 dogs. Initially the huskies performed well but as the journey progressed they rapidly deteriorated. Traditionally, Eskimos drive their dogs very hard but Scott was reluctant to use the whip. Soon the men were pulling the sledges, whilst the dogs trotted along next to them and sometimes even sat on board. The other two men on Scott's team, Ernest Shackleton and Bill Wilson, realised the necessity of feeding the sick dogs to the others, again standard practice amongst the Eskimos, but such was Scott's feeling for animals that he refused to do any of the killing himself. 'It is a moral cowardice of which I am heartily ashamed,' he later wrote. When all of the dogs finally died, Scott was relieved of what he had come to see as a burden:

'No more cheering and dragging in front, no more shouting and yelling behind, no more clearing of tangled traces, no more dismal stoppages and no more whip.'

ALASKAN SLEDGE DOG COMMANDS
✳ ✳ ✳

'Mush, mush', usually the signal in Hollywood movies for dogs to get going, is a corruption of the French verb 'marcher'. In fact, it is not really used by dog drivers because 'mush' is considered too soft a sound to be commanding.

Hike up!	Start	
Whoa!	Stop	
Gee!	Move right	
Haw!	Move left	
Easy!	Slow down	

This experience convinced Scott and Shackleton that sledge dogs were no real use in the Antarctic. They were wrong; much of the success of Roald Amundsen's expedition to the South Pole was based on his careful use of sledge dogs. Huskies continued to be used in Antarctica until the early 1990s, when they were banned under a new Protocol on Environmental Protection. At the time it was feared that canine distemper was spreading from dogs to seals. The Antarctic's huskies headed back north.

Camels

Camels originated in Asia and then spread to the Middle East. In the nineteenth century, they were exported to Australia and North America to be used as pack animals. Camels are uniquely adapted to life in the desert: their long eyelashes protect them from wind-blown sand; their wide, padded feet allow them to walk comfortably on steep dunes and their thick coat keeps the cold at bay at night, and the sun at bay during the day.

The camel's most distinctive feature is its hump: Arabian camels have one and Asian camels have two. It's a common myth that a camel stores water in its hump; in fact, it is a fat store which can contain up to 100 pounds (45 kilograms). The hump will visibly shrink if a camel has to go hungry for many days and will eventually collapse. A camel also has a very sophisticated water-rationing mechanism. Whereas humans need to maintain a constant core temperature of 98.4°F (37°C), camels can survive with a reduced temperature, without any ill effects. A camel's core temperature can drop to 98.4°F (37°C) at night and then rise to 105°F (41°C) in the heat of the day, before it starts sweating. When water is scarce, camels concentrate their urine, thereby losing less liquid; they are also able to draw water out of their tissue into their bloodstream, thus stopping their blood from becoming excessively thick. When they do find water, a camel can drink over 20 gallons (95 litres) in ten minutes and quickly restore its dehydrated tissue.

Freya Stark described the camel as the most perfectly adapted creature to its environment that she had ever met, but Sir Richard Burton was not quite so complimentary:

The so-called 'generous animal', the 'patient camel', whose endurance has been greatly exaggerated is a peevish, ill-conditioned beast – one of the most cross-grained, vile-tempered, and antipathetic that domestication knows. When very young it is cold, grave, and awkward; when adult, vicious and ungovernable, in some cases even dangerous; when old it is fractious and grumbling, sullen, vindictive, and cold-blooded.

EXPEDITION PETS

Exploration can be a lonely business, so it is not surprising that pets frequently appear in expedition narratives. It is a risky pastime though, for pets as well as their masters.

Nansen was the ship's cat on board the *Belgica*, which went to the Arctic in 1897. The *Belgica* was trapped by the ice pack for over a year. Everyone on board became ill and depressed, Nansen the cat included. He spent a month in a listless stupor punctuated by episodes of uncharacteristic bad temper, before he disappeared never to be seen again.

Mrs Chippy was a cat taken south by Tom McNeish, one of the carpenters on *The Endurance*, the ship taken to the Antarctic by Ernest Shackleton in 1914. When *The Endurance* had to be abandoned, Shackleton gave the order that Mrs Chippy and all the expedition's dogs should be put down. McNeish never forgave Shackleton for having his cat killed.

Tschingel was a dog owned by the Victorian climber and Alpine historian William Coolidge. She accompanied her master on many Alpine outings and made eleven first ascents. At one stage she was said to have been nominated for membership of the Alpine Club but rejected on grounds that she was a bitch; at that point the club was men only. Tschingel had 34 pups.

Toby was a pet pig taken to the Antarctic in 1903 by the French explorer Jean-Baptiste Charcot. Toby didn't last the winter: he died after snaffling his way through a bucket of fish, which unfortunately still contained the hooks on which they had been caught.

Titina was a fox terrier owned by the Italian explorer and airship designer, Umberto Nobile. He took her on two trips to the North Pole in 1926 and 1928. Titina also had the distinction of being one of the few dogs to urinate on the White house floor after Nobile was invited in to see President Coolidge in the Oval office. When Titina died in 1938, she was stuffed and put on display at the Italian Airforce museum in Lake Bracciano. She is still there.

ANIMALS THAT CAN EAT YOU

Whilst pack animals and pets are a useful addition to many expeditions, wild animals are a potential hazard. In the Arctic polar bears are the problem; in the Antarctic beware of killer whales; in Africa and South America predators are legion. Usually man has the upper hand but you can't be too careful . . .

HOW TO DEAL WITH A
RATTLESNAKE BITE
* * *

1. Do not suck the poison out – you will contaminate the wound with bacteria
2. Do not cut into the wound – this will damage the tissue around the bite
3. Do not put on a tourniquet unless you are sure of what you are doing
4. Do wash and clean the wound
5. Do try to immobilise the area around the bite with a splint
6. Do head for the nearest medical outpost as fast as possible

HOW TO AVOID BEING
ATTACKED BY A SHARK
* * *

1. Be extra careful at dusk and night, a shark's favourite feeding time
2. Do not enter the water if you have an open wound
3. Take special care if you see schools of baitfish or any pilot fish – this is a sign that sharks are about
4. If you realise you are being stalked by a shark, move slowly towards the shore or your boat
5. If you are diving with a partner and sight a shark, it will be easier to defend yourself if you tread water, standing back to back. Sharks can be put off by giving them a blow to the nose or the head with an oxygen cylinder

How to deal with a charging lion

Remember, old and wounded animals are more liable to attack, as are mothers with cubs. Be careful of tall grass – lions love to hide in it. If you see groups of zebras or giraffes staring at a thicket of grass, it may be that there is a lion lurking inside it. If you see animals which seem reluctant to go to a watering hole, this may be another sign that lions are about. Lions do most of their hunting at night, so in lion country it is important to keep a fire going in the evening and always to have someone keeping watch.

If you are confronted by a lion do not run away because it will charge; back away slowly. When a lion flattens it ears and swings its tail stiffly up and down this is a sign that it is angry and about to attack; if its ears are upright and its tail is twitching more randomly, then it may just be showing signs of excitement. Unless you've got a gun and are a sure shot, the best way to defend yourself is to stand still, take on a posture of aggressive defiance and stare the lion down when it runs at you. This obviously requires a lot of nerve, especially when the lion may charge more than once; each time it will stop a few feet in front of you and growl before turning away. In the Kalahari, the Bushmen believe that you should do more than just stare the lion down: they advocate making a lot of noise and throwing sticks and rocks at it.

Even if you have a gun it is not so easy to kill a lion, as the British explorer David Livingstone discovered in 1844. He was living at a mission house in an isolated valley in South Africa. When one day some local people told him that a lion was preying on their sheep, he set off to hunt it down. When Livingstone came face to face with the lion, he gave it both barrels of his shotgun but the lion kept coming. It seized Livingstone by the arm and shook him 'as a terrier does a rat' and then went on to attack two natives before it finally dropped dead. Livingstone never wholly recovered from the wounds the lion inflicted.

How to survive a charging elephant

A charging elephant is a truly terrifying sight: a fully grown male can weigh up to 13,200 pounds (6,000 kilograms) and stand 11 feet (3 metres) high. Beware of herds with small calves, and of any animals which have damaged tusks or look as if they have been hunted in the past. Elephants have poor eyesight but a very highly developed sense of smell and good hearing; try to stay downwind of them.

Like lions, elephants may make mock charges: if they come at you with their ears spread out, it's often a sign that this is for show; if their ears are flattened back then the charge is for real. If it is a mock charge, then it is better to stand your ground and wait for them to stop in front of you. Elephants can sprint up to 25 miles/hour (40 kilometres/hour) so the only way to escape is to start running well before they charge. If you do run and they appear to be catching up, move sharply to the right or left on the downwind side; if you're lucky the elephant will often rush past you. An elephant may make an initial attack with its trunk and then attempt to gore its victim with its tusks or simply crush it to death. If you're down on the ground, your only hope is a well-armed rescuer. Large-bore rifles are more likely to bring down an elephant than those of a lower calibre, but they are difficult to use because of their ferocious recoil. Be aware of any other elephants nearby: a lot of the fatalities which occur during elephant hunting are caused, not by the target elephant, but by another member of the herd which attacks from the side or the back.

How to avoid a bear attack

Bears generally avoid confrontations with humans, so in bear country you should try to make a lot of noise in order to warn them of your presence. Some people shout every 50 feet (150 metres), while others wear 'bear bells'. If you are confronted by a bear, you should hold your hands above your head in order to appear larger. Then back away slowly whilst talking to the bear; if you run, it'll chase you and over small distances a bear can move fast. Pepper spray can halt a bear, but it can also make it angrier. As you are backing away, leave a trail of items behind you. Bears are by nature very curious and easy to distract.

A polar bear, though, is very dangerous company. It is the world's largest predator and can weigh up to 1,000 pounds (450 kilograms). It has better eyesight and a keener sense of smell than a brown bear and has an acute sense of hearing. Whereas a brown bear lives mainly off vegetation, a polar bear is omnivorous and has very sharp teeth, designed to tear off large chunks of meat from its prey. To stop a polar bear at close quarters you need a large-calibre gun but it can sometimes be frightened off with signal pistols and loud sirens.

Most encounters involve solitary bears or a mother with a few cubs, but at certain times of the year polar bears leave the ice en masse and head for the nearest land. In 1969 the British explorer Wally Herbert ended his epic trans-Arctic expedition off the coast of Spitsbergen. As he and his men headed down towards their support ship HMS *Endurance*, they found themselves repeatedly menaced by polar bears. It was particularly worrying because they were carrying only quite low-calibre rifles. The British team were attacked so many times that Herbert started to worry that his ammunition might run out. Throwing ice-axes and boots at the bears didn't seem to make any impact. As they drifted around on an ice floe in seeming circles, their concerns grew but then after a few days they managed to get on to much firmer ice and escape from the area.

But...

In general humans are usually more dangerous to bears and other wild animals than vice versa. There are many more examples of explorers who have survived on polar bear meat, than polar bears who have tasted explorer flesh (or living flesh at least).

ANIMALS YOU CAN EAT

For anyone going on a long expedition, fresh meat can be a very valuable addition to his or her diet. Today there are countless rules and regulations about hunting game, but in the nineteenth and early twentieth century guns and ammunition were a standard part of expedition equipment.

Seal

Seal is still a staple item of the Eskimo diet. Traditionally, it is eaten raw. Every part of a seal has its own value: the eyes and the blubber are good sources of Vitamin A, the tongue and the heart are very tender and the brain is considered a delicacy. Seal meat has often been a life-saver for polar expeditions. In 1914, a party of Russian sailors found themselves fleeing across the Arctic after their ship became trapped in the ice. Whenever they could, they ate seal. In his journal Valerian Albanov, the leader of the party, noted that the meat from seals that lived off crustaceans tasted like venison, whereas other seal meat tasted much fishier. He agreed with the Eskimos on the delicate taste of seals' brains and added that the front flippers were reminiscent of calves' feet. Today the Inuit of Northern Canada still hunt seal to supplement their diet; there is also quite a lot of larger-scale commercial sealing in Newfoundland but most of the meat is turned into silage. In Asia there is a small market for seals' penises; they are sold at high prices for their supposed aphrodisiac qualities.

Penguin
Penguins are friendly, docile creatures which are painfully easy to hunt. They have a substantial inner lining of fat which needs to be removed before cooking and they taste similar to seal. Penguin liver is a noted delicacy and there are lots of references to it in Antarctic diaries. Scott's men ate penguin for their Christmas dinner in 1911. Today it is off the menu: under the Antarctic treaty of 1990, you are no longer allowed to kill a native animal except in 'cases of extreme emergency involving the possible loss of human life'.

Dog
Amongst the British there was a deeply held aversion to eating dogs but other explorers didn't have such qualms. The Norwegian, Roald Amundsen, ate dog meat on his trip to the South Pole. He noted that dog cutlets tasted particularly good. Whilst Captain Cook was exploring the coast of Australia he was offered roast dog by a local chief. Cook declared that he had never eaten sweeter meat.

Albatross

The first time I killed one I felt like a murderer, the second time a little less bad, and after that I just thought what a fine meal they would make, and what a glorious feed the first had been.

FRANK WORSLEY, CAPTAIN OF THE ENDURANCE,
SHACKLETON'S LOST SHIP

There are a lot of myths and superstitions about albatrosses, but that hasn't kept them out of the stew pot. They can live to 60 years of age and, with their huge wings, are able to travel enormous distances gliding on the wind. Their flesh is white and strong tasting. The Maoris of New Zealand used to regard albatross meat as a great delicacy; and even though they were said to carry the souls of dead sailors, they were still shot and trapped by the crews of whaling ships. When Shackleton and his men arrived at South Georgia after a desperate 14-day journey in a small boat, they landed near to a colony of albatrosses. After suffering initial pangs of conscience, the starving explorers soon tucked in. At one point Shackleton and Worsley mooted the idea of setting up a business to sell the chicks to the epicures of London and New York before they remembered that there were international conventions which forbade the killing of albatrosses. At that point though, as Worsley later wrote, they were a law unto themselves.

Polar bear
Polar bear meat was once a significant part of the Eskimo diet, though bears were not nearly as numerous as seals and much harder to kill. Peter Freuchen the Danish explorer, who lived for many years in Thule in Greenland, wrote how at certain times of the year they would kill so many bears that he and his Eskimo wife would get bored of the taste. Like seal, bear meat can be stewed but is said to taste best when raw and frozen.

Camel

A healthy camel can weigh up to 1,400 pounds (650 kilograms) and provide a lot of meat. Camel tastes like coarse beef; the younger the animal, the more tender it is. In some countries the hump is considered a delicacy and is reported to be at its best when eaten raw and still warm. Camel milk was once an important source of food for the Bedouin of Arabia, so much so that they used to slaughter the males when young and always travel with females. The camel's South American cousin the llama is also widely eaten; it gives good-quality meat, high in protein and low in fat.

Kangaroo

Kangaroos have been eaten by the Australian Aborigines for centuries. The first Westerners to try them were probably the crew of Captain Cook's ship *The Endeavour*, which dropped anchor off Australia in 1770. Initially the crew was confused by their first sight of kangaroos: Cook described them as strange mouse-coloured creatures which looked like greyhounds but seemed to jump like hares or deer. After a fortnight, they had shot and skinned their first animal and learnt its native name, 'Kangooroo' or 'Kanguru'. Kangaroo meat is low in fat and best roasted or lightly fried. At one time it was banned in some states of Australia because it was thought to be worm ridden; today it is widely eaten, though a lot of kangaroo meat ends up as pet food.

Food

The most important thing, is that there is some.

TOM BOURDILLON, MEMBER OF THE
BRITISH 1953 EVEREST TEAM

All explorers and mountaineers need an adequate supply of food to keep them healthy and it's a logistical challenge to make this happen. When an expedition goes into the field for several years, it's virtually impossible for it to carry all of its food supplies, so alternative sources must be found. This usually means hunting for meat and fish or buying food from local people. The less an expedition carries, the faster it can move and the fewer porters or pack animals it needs. On big expeditions, this can be a serious issue. In 1977 a Japanese expedition went to K2 in the Karakoram Range with 52 climbers; they needed 1,500 porters to carry their food and equipment.

Robert E. Peary's recipe for Pemmican
66% lean beef, dried and ground to a fine powder
33% beef fat
A little sugar
A few raisins

Pemmican used to be the staple food of nineteenth- and twentieth-century explorers. Whether you were in the Arctic, the Himalayas or the wilds of South America there was usually a tin of pemmican lurking somewhere nearby. Pemmican was invented by Native Americans and then adopted by white fur-trappers. High in protein and fat, it could be preserved in tins or moulded into balls. Usually pemmican was added to boiling water to make a soup, or 'hoosh' as Shackleton called it, but it could be eaten raw. The problem with pemmican was that, because it contained no carbohydrate or roughage, it tended to cause constipation. The other problem was that if the cooking and canning were not done carefully, tins of pemmican frequently went mouldy. It's remarkable the number of explorers who survived on a diet of mouldy pemmican; in extreme situations the stomach seems to be able to tolerate a lot.

A rich and varied diet is for people who have no work to do.

ROALD AMUNDSEN, THE FIRST MAN TO
THE SOUTH POLE

There are two basic schools of thought when it comes to expedition food: one is that it should all be kept as simple as possible in order to keep the weight down, and the other is that at least a few luxuries are essential for the sake of morale. General Charles Armitage Bruce, the leader of British Everest expedition in 1922, was a firm advocate of food as a morale booster. In addition to the usual pemmican stews and sardines, he treated his team to champagne, tinned quail, foie gras and caviar. This gourmet approach continued into the 1930s for British Everest expeditions until Eric Shipton was asked to be leader in 1935. His philosophy was entirely different: he disliked tinned food and thought luxuries were an unnecessary encumbrance. On the march in to the mountain, his expedition ate mainly local food bought from villages along the way. Shipton's party of seven was able to put away 140 eggs in a single day and was happy to eat the local yak butter and cook with local flour. When they started climbing, Shipton opted for a very simple pemmican-based diet after consulting a nutritionist in London. Though it could be argued that frequently he was just too spartan, Shipton certainly took food very seriously. He observed that in cold weather a lot of climbers crave fat, but at high altitude it is very unpalatable and difficult to digest. He also understood that mountaineers tended to be lazy about cooking when they got really high up, but that if they were to perform well, they had to force themselves to eat. In the end Shipton came to the conclusion that the best thing to do was to adopt a carbohydrate-based diet similar to the local Tibetans; it was easy to digest and easy to carry.

When expeditions are not planned well or hit by an unforeseen disaster, food rapidly becomes an obsession.

JOHN HERRON'S DIARY

John Herron was the steward on the *Polaris*, an American ship which went to the Arctic in 1871; in a very bizarre incident, half of the crew were abandoned on an ice floe whilst the other half steamed away with a drunken captain at the helm. Herron was amongst the abandoned men. They were not rescued for eight months; food was a constant preoccupation.

Oct 31st	We have been living very poorly so as to make our provisions last six months.
Nov 16th	We have nothing to feed our dogs on; they got at our provisions today; we shot five, leaving four.

Dec 2nd	Boiled some seal-skin today and ate it – blubber, hair and tough skin.
Dec 6th	A poor fox was devoured today by seven of the men who liked it; they had a mouthful each for their share; I did not think it worthwhile myself to commence with so small an allowance, so I did not try Mr Fox.
Dec 7th	All in good health. The only thing that troubles us is hunger; that is very severe. We feel sometimes as though we could eat each other.
Jan 16th	Hans caught a seal today; thank God! for we were very weak. I am afraid I have a touch of the scurvy. A little raw meat will drive it out, I hope.
Feb 26th	We are coming down on our provisions one half; that is as low as we can come and keep life, and will be a few ounces a day.
Mar 27th	A very agreeable surprise tonight while at supper. A bear came to the hut. Of course, he died; we buried him in the snow until morning.
Mar 28th	Skinned and cut up the bear; he is a fine young one, very tender and fat, weighing I should say 700 or 800 lbs. We are making some sausages from him, which are very good I think. I think it is the sweetest and tenderest meat I ever ate. The fat cuts like gelatine.

Their ice floe breaks up and they have to pack into a lifeboat.

April 1st	Got underway the boat taking in water. Loaded too deep. Threw overboard one hundred pounds of meat; must throw away all our possessions.
April 16th	We have but few days provisions left. The only thing that troubles me is the thought of cannibalism. It is a fearful thought, but may as well be looked boldly in the face as otherwise. If such things are to happen we must submit. May God save us!
April 30th	Glorious sight when fog broke; a steamer close to us. She sees us and bears down on us. We are saved thank God!

FOOD FIGHTS

Food can make members of an expedition very selfish. John Hunt was leader of the 1953 British Everest expedition. In his account of the expedition he describes an incident when, high up on the South Col, he found a tin of tuna fish which had been left by a previous expedition. Rather than share it with the other members of the team, he snuck off to a small tent and ate it all himself. Appetite and animal instincts he noted, can make a climber very unsocial.

Ernest Shackleton's diary for his expedition to the South Pole in 1908 is full of elaborate food fantasies. He and his three companions stopped just 100 miles (161 kilometres) from the Pole and spent most of the return trip on half rations, dreaming about food. In one telling moment they found a few pieces of biscuit and some chocolate left in the snow by an earlier party. When they drew lots for their find, Shackleton didn't get the piece that he wanted. The normally selfless leader was not pleased: 'A curious unreasoning anger took possession of me for a moment at my bad luck. It showed how primitive we had become and how much the question of even a morsel of food affects our judgement.'

Three years later another party of British explorers was starving in the Antarctic. They had come south with Robert Falcon Scott; whilst he headed for the South Pole, Victor Campbell and a party of five men were sent to explore the land to the north west of Scott's base. In May 1912 they lay huddled in an ice-cave on Inexpressible Island, having been forced to spend a second winter in the field because their relief ship couldn't make it through the ice pack. They managed to lay up a food cache of 120 penguins and kill some seals, but the relentless diet of meat gave them dysentery and made their urine so acidic that several of them became incontinent. They all dreamt a lot about food and for entertainment would recount their dreams to each other. For most of the men the stories always ended in frustration: they would reach a grocer's shop, only to find that it was closed or lay a table only to wake up before they could eat anything. Two of the men, though, had dreams that usually ended successfully and delighted in recounting the elaborate meals that they had eaten, whilst asleep. This didn't go down well with the other men and, in their annoyance, they went so far as to suggest that the happy dreamers should have their real-life rations reduced because they were eating so well in their sleep. Sweet dreams indeed.

In Extremis

Wilfred Thesiger, the British explorer who made two crossings of the Empty Quarter in Arabia always ate and drank the same food as his Bedouin guides. However, he baulked at their most extreme practice: if they found water which was too brackish for human consumption, the Bedouin would make their camels drink it, put a stick down their throats and then drink the vomit.

Heinrich Barth was a German scientist who became famous for his exploration of North Africa and his journey to Timbuktu. At the end of a solo excursion to climb Mount Idinen in the Sahara, he ran out of water and collapsed. In order to quench his thirst, he wrote how he 'sucked

a little of my blood until I became senseless'. He didn't elaborate on where he got the blood from but fortunately he was rescued on the following day by a Tarki tribesman who had spotted his footsteps in the sandy ground.

Thor Heyerdahl, the man behind the *Kon-Tiki* expedition, was a soldier in the Free Norwegian Forces during the war. Though he never experienced it himself, he heard stories of shipwrecked sailors who had quenched their thirst by extracting the liquid out of fish. They found that if they cut into the side of a large fresh fish, liquid would collect in the exposed flesh due to lymphatic drainage. Alternatively, a fish could be chopped into small pieces, wrapped in cloth and then the moisture wrung out.

John Franklin, the Victorian explorer, became famous, after publishing a graphic account of his Arctic expedition in the 1820s, as the 'man who ate his boots'. When his party ran out of food they resorted to eating roasted leather taken from their boots and old animal skins. To this they added a soup made from 'tripe-de-roche', a type of lichen which they found growing on rocks. Franklin wasn't the only explorer to resort to eating his clothing. In 1903 the American adventurer, Leonidas Hubbard, embarked on a disastrous trip to Labrador in Northern Canada. His diary concluded with the hapless explorer looking forward to eating his moccasins and his cow-hide mittens. He did not survive for long.

Adolphus Greely led an American expedition to the Arctic in the 1880s. After spending two years at a remote fort, he and his men headed south when their relief ship didn't turn up. It soon became a starvation march with the men forced to survive on old pieces of seal skin. When rescuers finally arrived and found the men in a makeshift camp, they exhumed several bodies only to find that, though the faces had been left intact, chunks of flesh had been cut away.

THE HUMAN FACTOR

Understanding how to survive in a particular climate is vital for an expedition to succeed but the objective hazards of climate and terrain are not the only issues that an explorer has to deal with. The human factors of expedition life can be just as complicated and just as demanding. Leadership, teamwork, dealing with strangers, facing up to solitude: these are just some of the issues that are treated in the next chapter.

Solitude

Empathy

Fear

Boost

Bo

ringo

Lea

Desertion

Leader

Mutiny

Cheer

I

Charisma

Na

Player

3

Getting Along

FREYA STARK'S TEN QUALITIES
NEEDED BY A TRAVELLER

In the public mind, explorers spend most of their time battling with the elements, but the human struggles, which are very much a part of expedition life, are equally demanding. In order to succeed, the physical stresses of climate and environment have to be dealt with, but so too do all the mental pressures of working in the field. Some of the great explorers preferred to travel in the sole company of local guides but, for most, exploration was a collective experience whose success usually depended on their leadership skills and the qualities of the teams they put together at home.

LEADERSHIP

A good leader is...
A good organiser
A good communicator
A good listener
A team builder

He or she is...
Flexible
Positive
Unflappable
Selfless
Confident

But... what's easy on the page is much harder in real life.

For some people leadership is all about giving orders. Robert Peary, the American polar explorer, was the archetypal strong man. He was physically and mentally tough, and as hard on himself as he was on others. He spent 20 years in the Arctic, exploring the unmapped coast of Greenland and making repeated attempts to reach the North Pole. By the end he had honed his approach to a fine art: he would take a ship as far north as he could, round up the local Eskimos and then spend the winter getting prepared. Then in the spring, he would set off for the Pole. The other members of the team would do all of the early work, breaking the trail and moving up the heavy supplies, before Peary's small, well-rested party leap-frogged over them and made a dash for the Pole. It was a method of working which could only have been devised by an authoritarian leader. Robert Peary was not known for his charm or his man-management skills; he frequently fell out with colleagues and, as his exploration career progressed, he increasingly preferred to travel with Eskimo assistants rather than American companions.

Eric Shipton, the British explorer and mountaineer, was known for a very different style of leadership. He preferred small expeditions and despised the glory hunting that characterised much Himalayan mountaineering. In the 1930s, Shipton and his friend, Bill Tilman, used to boast that they could organise an expedition on the back of an envelope in half an hour. They were an odd couple: both instinctively Spartan in their attitude to travel, for many years they didn't even use their Christian names when addressing each other. In 1937 Shipton and Tilman put together an expedition to the Karakoram Range in Pakistan. They took a small band of Sherpa porters and two British surveyors, Michael Spender and John Auden, both brothers of famous poets. Their aim was to visit the unexplored glaciers and mountain passes around the Shaksgam River, to survey the region and revel in the simple pleasure of filling in a blank on the map. Everything was very informal and low key: the expedition was financed on a shoestring; they ate very simple food and kept their equipment to the absolute minimum. After several weeks together, Shipton realised that because there was so much new ground to cover it would be better to split the party into three separate units, each to explore a particular area. When several weeks later the men reconvened in Kashmir, they were thrilled at what each had achieved and started planning a return trip. This modest, low-key approach won Shipton many armchair fans and made him very attractive to fellow climbers and explorers.

There were limitations, though. In 1952 Eric Shipton was sacked from the leadership of the forthcoming British Everest expedition and replaced by an army officer, John Hunt. Officially it was a resignation, but the organising committee made it very clear that they didn't want Shipton to be the leader. Earlier that year, he had taken an expedition to a much lower Himalayan peak, Cho Oyo. It had been intended as a training expedition for Everest, but Shipton's team failed to reach the summit and came back in a slightly shambolic fashion. This convinced the Everest organising committee that Shipton just didn't suit big expeditions. He was a true explorer who loved nothing more than to discover new lands; getting his name in the record books held no great attraction. Shipton had leadership thrust upon him because he was an ambitious and effective mountaineer, but he wasn't naturally drawn to it.

Shipton's *laissez-faire* approach to leadership and Peary's author-itarianism are two extremes between which most leaders fall. It isn't easy to strike the right balance between giving orders and encouraging teamwork and initiative. With this in mind, it is interesting to look at three expeditions to K2 between 1939 and 1954; they were very different but each one provides lessons in leadership.

Three Styles of Leadership on K2

K2 is the second highest mountain in the world. Its nicknames tell you a lot about it: 'the Savage mountain', 'the Killer mountain', 'the Moun-taineer's mountain'. It lies at the heart of the Karakoram Range on the border of China and Pakistan. It's 800 feet (240 metres) lower than Everest but most climbers regard it as a tougher proposition and, unlike Everest, there are no commercially guided expeditions on K2. Between 1954 and 2004, 240 climbers reached the summit and 60 died trying. The first serious attempt was made in the late 1930s by an American team led by Fritz Wiessener.

LEADING FROM THE FRONT: FRITZ WIESSENER

Fritz Wiessener was a German emigré who arrived in the United States in 1929; initially he came on business, but within a few years he decided to make America his home. He was a keen mountaineer and skier and he soon made a name for himself amongst American climbers. Wiessener was incredibly strong and very technically skilled; like other Germans and

Austrians of the 1930s, he was more willing to take risks than previous generations of climbers. Soon Fritz became involved in the American Alpine Club. He set up a joint German–US climbing expedition to Nanga Parbat in the Karakoram Range and made the headlines for first ascents of some of America's most challenging peaks. No one was surprised when he was offered the leadership of the 1939 American Alpine Club expedition to K2.

Wiessener did, however, have a lot difficulty in recruiting a strong team to accompany him. Ironically, many of the best young American climbers had gone on a reconnaissance expedition to K2 in the previous year and simply weren't available for the main event. Others were put off by Fritz's Teutonic manner and his reputation for driving his companions hard. Eventually Wiessener did manage to assemble a team but, as climbers, none of the others were in his league. The men bonded well on the long journey from the US to the foot of the mountain, but as soon as they got to K2, things started to unravel.

One climber, Chapell Cramner, came down with suspected pulmonary oedema and even though he recovered, it effectively put him out of action for the remainder of the expedition. Wiessener wasn't discouraged though, and he quickly started work on his 'siege' of K2. His plan was to set up a series of camps up the South East Ridge, gradually moving up the mountain until he was in a position to make an assault on the summit itself. Bad weather forced him to delay the schedule and gradually the rot started to set in.

The two youngest members of the team, Chapell Cramner and George Sheldon, were clearly intimidated by the sheer scale of K2. Fritz's second-in-command, the 47-year-old Tony Cromwell, considered himself too old to climb high and he didn't like Wiessener. The only man who really warmed to Fritz was Dudley Wolfe, a millionaire socialite, who really didn't have the experience to be on the expedition at all. Most of the other climbers felt his only qualification was his cheque book and told Wiessener that Wolfe shouldn't be allowed to go up to the high camps because he was such a novice.

Wiessener brushed aside the complaints; he was feeling strong and confident and he relished the challenge. If he couldn't lead the others up the mountain then at least he could reach the summit himself. In some ways, deciding to go ahead with the attempt was Fritz's first big mistake; with such a weak party, it might have been better to have scaled down his ambition and climbed one of the smaller peaks nearby. It was not surprising, though, that Wiessener didn't make this choice: here he was with a chance to make history, and he wasn't going to give up straight away.

After six weeks, Wiessener passed the highest point of the previous year's reconnaissance expedition. By now, he was more or less disengaged from the other men; they had effectively been turned into his support party. Most of them were down at base camp, wishing they were somewhere else. Mountaineering is by its very nature a waiting game and most climbers accept that on a big mountain there is always a lot of 'pit-time' when bad weather keeps them trapped in their sleeping bags. The problem on this expedition was that whilst Wiessener had his eyes on the summit, most of the others just wanted to go home; it wasn't the weather that was holding them back, it was Fritz.

By choosing to lead from the front, Wiessener fell out of touch with the rest of the team. In 1939 it wasn't possible to take a workable radio system on to a mountain like K2; apart from the occasional handwritten note, communication was minimal between Fritz and the other men. Wiessener didn't realise how unhappy they were, and they didn't follow the orders that he had given them. Wiessener thought that as he was climbing higher and higher up the mountain, they were continually re-stocking the camps below him and making sure that he would have a secure line of retreat for when he had to come back down. The reality was that most of the men were either unwilling or unable to go high, so increasingly they left the work to their Sherpa porters. Ironically, the only man who was able to cope with the altitude was Dudley Wolfe, the least competent mountaineer. He went with Wiessener all the way to his eighth camp; Wolfe stayed back, however, when Wiessener headed for the summit itself.

On their first attempt, Wiessener and Pasang Lama, his Sherpa climbing companion, came very close to the top but were forced to turn back because Pasang was superstitious about climbing at night. After an abortive second attempt they headed back down the mountain, aiming to pick up fresh supplies and climbing equipment. To Wiessener's shock he found that the camp immediately below him hadn't been resupplied at all and that its solitary occupant, Dudley Wolfe, had not seen anyone for three days. As Wiessener continued down the mountain, his shock turned to rage as one by one he found that the lower camps had been abandoned. How could this have happened? Why hadn't they followed his orders?

When finally he stumbled on to the glacier at the foot of the mountain, frost-bitten, parched and thoroughly worn out, he met his second-in-command Tony Cromwell searching for Fritz's dead body. Cromwell revealed that in the previous weeks none of the American climbers had felt strong enough to go up the mountain and had left the Sherpas to do all the work. One of the Sherpas had seen signs of an avalanche near Fritz's top camp and had told the others that Wiessener's party must have been swept away. The Sherpas then proceeded to strip the upper camps; the lower ones had already been emptied in anticipation of everyone returning home. Fritz flew into a rage and hoarsely accused the others of sabotage and attempted murder.

Eventually, Wiessener calmed down. He said that he wanted to go back and make another attempt on the summit but in reality he was physically and mentally exhausted. His immediate problem though was getting Dudley Wolfe down. Wiessener had left him high up on the mountain, at his own request. This was undoubtedly another mistake. Even if Wolfe didn't want to come down, Wiessener should have insisted because Wolfe was so inexperienced. In his defence, Wiessener hadn't intended to leave Dudley for long; when he left him he had no idea that he would have to go all the way to base camp to find the rest of the team. Whoever's fault it was, someone now had to go up and get Wolfe down. First Jack Durrance tried; he was the strongest of the remaining climbers but the altitude forced him back. Then four Sherpas went up, led by Pasang Kikuli, the head Sherpa. Only one Sherpa returned: the others were presumed to have been killed by an avalanche. Finally Wiessener himself made a desperate attempt to climb back up but he was all played out; he didn't get very far and saw no sign of anyone above. A thoroughly defeated team fled the mountain.

Back in America, the arguments that had simmered at base camp broke out into vicious hostility. Wiessener's deputy, Cromwell, accused him of recklessly endangering Dudley Wolfe's life by taking him too high. Well-known climbers entered the fray criticising Fritz's leadership: they said it was a mistake to break the party up and that Wiessener had seriously misjudged Dudley Wolfe's ability. Wiessener, in his turn, blamed the disaster on the failure of the other men to stick to his plan: if they had kept the camps well stocked with food and equipment, Wolfe would still be alive, he maintained. The arguments were never resolved and the scandal lingered on.

In the 1980s Wiessener was rediscovered by a new generation of climbers who hailed him as a great mountaineer. It is hard not to agree, but on K2 he was not a great leader. In the first instance, he was handicapped by a weak team. This wasn't his fault but it did inevitably push him towards leading from the front with all its attendant risks. When it came to Dudley Wolfe, at best Wiessener was optimistic, at worst he was negligent. Good communication is a prerequisite of good teamwork; by choosing to lead from the front, Wiessener made communication very difficult.

LEADING FROM THE MIDDLE:
CHARLES HOUSTON

With the recriminations over the 1939 'tragedy' still hanging in the air, in 1953 Charles Houston, another American climber, put together a new team to take on K2. Houston was one of America's most famous mountaineers: in the 1930s he'd made the first ascent of Mt Foraker in Alaska and had put together an Anglo–American expedition to Nanda Devi in India. He had some history with K2: in 1938, a year before Wiessener went out, Houston had led the American Alpine Club's reconnaissance expedition. It went very well; Houston's men had managed to find a workable route and had in fact climbed much higher than anyone had expected. Houston was critical of Wiessener's expedition in the following year; he too felt that Fritz had not been a good leader and was particularly angry about the deaths of the Sherpas at the end of the expedition.

Houston was able to put together a strong party. His second-in-command was Bob Bates, a talented climber and a close personal friend who was also a veteran of the 1938 reconnaissance expedition. The other five men were all young but very promising. Two of them were professional guides, one was a ski-instructor and the others had made difficult climbs in the United States. Above all, none of them were 'prima donnas'; Houston's stated ideal was to put together a group that would work together cohesively and have a temperament that could deal with all the stresses of high-altitude climbing. Having the right personality and good all-round skills were judged to be more important than technical brilliance. This thoughtful and rigorous approach to selecting the team was one of Houston's strengths as a leader. Unlike Wiessener, he wasn't willing to simply make do.

In Pakistan, Houston's team was joined by a British army officer, Tony Streather; initially, he was taken on as a liaison officer but he quickly became a part of the team. Local porters carried their equipment to the foot of K2 but they went no further. This expedition didn't include any Sherpas so they had to do all their own load carrying when they reached the mountain. Houston's team took the same route as Wiessener and adopted the same siege tactics, but there was a crucial difference: instead of splitting the party and leaving some men down at base camp, Houston's men went up the mountain en masse. After six weeks things were going so well that the climbers started to discuss who would make the summit bid. Houston had already decided that he wasn't going to take part in the final stage; he and Dee Molenaar were the only two men with young families and neither of them wanted to take the risk. The others were not so cautious; they had a secret ballot for who would be in the first pair. It was all looking good: here was a strong and experienced leader, surrounded by a cohesive, evenly balanced group. Victory seemed to be in sight.

Then the storm hit.

They were trapped in their tents for five days, lashed and battered by relentless winds. When they finally abated the men emerged from their tents, only for the youngest climber, Art Gilkey, to collapse with a severe pain in his left leg. He thought it was just cramp, but Houston, a doctor by trade, diagnosed it as thrombophlebitis: blood clots had formed in one of his legs cutting off the circulation. The risk was that one of the clots might become detached and cause a fatal pulmonary embolism. Houston knew straight away that there was only a small chance that Gilkey would get down alive, but he felt a tremendous sense of responsibility as the expedition leader. Everyone agreed: whatever happened they would have to make an attempt to save him.

Gilkey couldn't move by himself, so they wrapped him up as best they could in an improvised stretcher and started to descend. It was very difficult work and they made slow progress. Then suddenly, on a steep, icy slope, George Bell slipped off, pulling his partner behind him. Within a few seconds several ropes had become entangled and everyone was flying down a steep slope to their certain deaths. All bar one man: Pete Schoening. He had been waiting higher up the slope with a rope attached to Art Gilkey to belay, or protect, him. This rope was wrapped around Schoening's ice-axe which was wedged behind a rock. As the other members of the team slid down the slope, their ropes became entangled with each-other until all the weight came onto Schoening's rope. The fibres stretched and the ice axe wobbled but somehow he managed to stand firm and hold all six men until one by one they stopped in their tracks. It was an amazing feat which Schoening shrugged off modestly as an instinctive reaction. Art Gilkey did not survive though; he was swept away in an avalanche.

Everyone was severely shaken. Dee Molenaar had broken ribs and George Bell had frost-bitten hands and feet. Charles Houston had bashed his head on a rock and was suffering from concussion. It was touch and go as to whether the seven men would survive at all, but because they were such a strong and supportive team, and because Bob Bates was such a good second-in-command, they all managed to get off the mountain alive.

Charles Houston later described the experiences on K2 as an 'epiphany'; he felt huge guilt as the expedition leader and its doctor. It appeared to him that he had failed on both counts. Houston never climbed again, preferring to devote his time to medicine and scientific research. He wrote an authoritative book on altitude sickness which took many years to research. Houston was strong minded and well organised, but he was no autocrat; he and Bob Bates spent a lot of time choosing the team, and on the mountain Houston's democratic style made them all feel that they were in it together. When disaster struck, his team was able to work together very effectively. For many climbers Houston came to represent the model leader, and the fact that, 50 years on, the members of his team were all still in touch with each other was a testament to the strong bonds of friendship that had been forged between them on what they dubbed 'The Savage Mountain'.

LEADING FROM THE BOTTOM:
ARDITO DESIO

In 1954, an Italian team led by Ardito Desio finally made the first ascent of K2. This was a very different affair from the two previous expeditions; unlike Houston and Wiessener, Desio had never done any serious climbing. He was a 56-year-old geography professor from Milan and he saw himself first and foremost as a scientist. When it came to K2, he was absolutely determined to succeed. He told the press that his goal was not to make an attempt on the mountain, but to 'conquer' it. Italy had a strong claim: the first proper reconnaissance of K2 had been made in 1909 by an expedition led by the renowned explorer, the Duke of Abruzzi. Desio himself had seen the mountain for the first time in 1929, on an expedition to the Karakoram Range lead by Abruzzi's nephew, the Duke of Spoleto.

Desio did his preparation well. He scoured Europe for the best equipment available and talked to Fritz Wiessener about his experiences in 1939 and Pete Schoening and Charles Houston about their experiences in 1953. He advertised throughout Italy for alpinists to take part in his expedition and put candidates through a rigorous selection process, drafting in officers from the Italian army and hordes of medical men to help select the best team. Before they left Italy, Desio gave the men a small illustrated guide to K2, packed with details of where previous expeditions had put their camps and lists of the food left behind by Houston's team. When they got to Pakistan, Desio organised the first aerial reconnaissance of K2 before hiring an army of porters to transport in several tons equipment to the foot of the mountain. No one could fault his logistical skills or the huge effort that he put into the expedition. However, though Desio was an excellent organiser, he was not a great man-manager. He saw himself as a general at the head of a military operation; his version of military being all about discipline and subservience, rather than initiative and pragmatism.

When the Italians reached K2, Desio stayed at the bottom, dishing out orders over the walkie-talkie or sending up handwritten messages, urging the men to conquer the mountain for Italy. He brought along all the accounts of previous expeditions and was constantly referring to them. When things were going badly, he warned the team that they would bring shame on their country if they didn't reach the summit. Not surprisingly, the climbers didn't pay too much attention to Desio's missives.

After barely three weeks, tragedy struck K2 again when Mario Puchoz died of pneumonia. It was a hard blow and it took almost a week to get his body off the mountain; Puchoz was buried next to a memorial that Houston's team had put up to Art Gilkey. Then the Italians started up again. After two more weeks and a huge effort, Lino Lacedelli and Achille Compagnoni finally managed to reach the summit – despite their oxygen having already run out. A few days later, the news was telegraphed back to Italy. Desio stayed on in the Karakoram to run an extensive programme of research with a team of Italian scientists. His final message to the climbing team was to warn them not to let the success go to their heads when they returned home.

Initially the expedition was very well received in Italy but before long things started to go sour. Desio fell out with the Italian Alpine Club over who should hold on to a trophy awarded to the team; Achille Compagnoni sued the makers of the official expedition film for the frost-bite that he had suffered when filming with a cine camera on the summit. One of the younger members of the team, Walter Bonatti, accused Compagnoni and Lacedelli of abandoning him on the night before their summit attempt. He and a Pakistani porter had come up to their camp with spare bottles of oxygen, and had ended up spending a freezing night out in the cold. Bonatti's accusations sparked a bitter row, several books and dozens of articles in the press.

Desio *had* won the day: he had succeeded where the great leader Charles Houston and the brilliant climber Fritz Wiessener had failed. Ardito Desio was far from a model leader though, and it's hard to imagine that many of the team would have been willing to go on another expedition with him. His organisational skills and his intense will to succeed had undoubtedly laid the foundations for a successful ascent, but the scandals and controversies which came afterwards were hardly surprising in view of the competitive way in which the expedition had been set up.

Together these three expeditions show that there are no hard-and-fast rules: good leaders can fail and poor leaders can succeed. Arguably Charles Houston was the best leader but his expedition in 1953 was the least successful, if success is measured by how high his team got on the mountain. If you look at the totality of the experience, considering what happened afterwards as well as what happened during a particular expedition, good leadership does have an impact in the longer term. Several men on Houston's expedition went on to have distinguished mountaineering careers, but of the Italian team only the embittered Walter Bonatti went on to make a name for himself.

TEAMWORK

Good teamwork requires:
Patience
Consideration
Good humour
Discipline
Self-sacrifice
Empathy

Like good leadership, the ideals are often hard to attain.

Today teamwork is not quite such an important issue in mountaineering and general exploration. The era of the large expedition is over, for the moment at least. Modern Himalayan mountaineering seems to be divided into fast, lightweight 'alpine' ascents by small parties and large commercial expeditions which don't require the same level of camaraderie and mutual support. The Arctic and the Antarctic are full of one- and two-person expeditions slugging it out for whatever fame there is left, whilst bookshops are full of tales of solitary exploits in Africa and South America.

Historically, though, teamwork was central to the success of many expeditions. Obviously, leaders play a major role in creating an environment where everyone works together, but team spirit can't simply be imposed; it has to come from within. Empathy and self-sacrifice are at its core; for an expedition to work, individual members have to be tolerant of each other and must be willing to put the good of the team ahead of their own personal ambitions. Again, it is interesting to look at some expeditions from the 'Golden Age' of Himalayan climbing, to find lessons in teamwork.

TEAMWORK ON ANNAPURNA: MAURICE HERZOG'S PARTY, 1950

One of the most successful mountaineering books of all time, *Annapurna*, by Maurice Herzog, told the story of how in 1950 a French team made the awe-inspiring ascent of a mountain in the eastern Himalaya. It turned its author, Maurice Herzog, into an international star: not only had he led the expedition but he had reached the summit himself, only to return with appalling injuries. The recent publication of several other books, however, has revealed just how much a story of teamwork the whole expedition was. Whilst Herzog still deserves a lot of credit it is now clear that he wouldn't have been able to make the ascent, and he certainly would never have been able to get down, without huge personal sacrifices by the other men in the team.

The expedition started very strangely, when at the offices of the French Alpine Club in Paris all the members of the team were asked to swear an oath of allegiance to the leader, committing them to do everything Herzog commanded. It wasn't uncommon for team members to sign contracts outlining their role and their wages, or assigning copyright to the expedition organiser, but this was something different. The expedition was being promoted as a huge national enterprise for the glory of war-battered

France. Of the nine men in the team, three of them were elite mountain guides from Chamonix, the centre of Alpinism in France: Louis Lachenal, Gaston Rebuffat and Lionel Terray. They would all play major roles in the expedition and their self-sacrifice would be an integral part of Herzog's triumph. They took the oaths but they thought it was all very odd.

Once Herzog's party reached Nepal, the first problem was *finding* Annapurna: there were no reliable maps of the area and no other Western mountaineers had ever even set foot on it. It took several weeks simply to locate it and many more days to find a route that was workable. In the back of everyone's mind was the oncoming monsoon, whose heavy snow-falls would put an end to any hope of making an ascent. By the beginning of June, however, they were high up the mountain, preparing to make an attempt on the summit. The first party would be Herzog and Lachenal, the second summit pair would be Rebuffat and Terray. Terray had forfeited his chance to be in the first pair after tiring himself out, restocking the middle camps. It was a vital role but one which the less experienced members of the team were unable to carry out.

On 3 June 1950 Lachenal and Herzog did indeed reach the summit, but only at huge personal cost. Lachenal had not wanted to carry on; as they were ascending, he could feel his feet becoming more and more numb and he didn't want to risk frost-bite. As an experienced mountain guide he knew only too well the havoc it could wreak and he was worried that if he lost his toes or fingers, his livelihood would be threatened. Herzog, however, said that he wanted to continue and Lachenal realised that if his leader went alone, there would be very little chance that he would return. So, like a dutiful mountain guide, Lachenal carried on. When the two men came down to their high camp, they were greeted by Rebuffat and Terray, who were poised to make their own attempt on the summit. But it was not to be: they realised straight away that Herzog and Lachenal had severe frost-bite and needed to be evacuated as soon as possible. Lachenal's boots literally had to be cut off him; Herzog's were as hard as wood and his feet were frozen solid.

On the following morning they began their descent. Selflessly, Terray swapped his boots for Lachenal's, risking severe frost-bite himself. The four men battled their way through a storm and then spent a brutal night, huddled together in a crevasse with one sleeping bag between them. By now Rebuffat and Terray were snowblind and Herzog was in the throes of despair. He asked to be left behind but the other men would hear nothing of it and lifted him on to his feet. Finally, their luck turned; they found two other members of the team who had come up to help and soon they were being helped down the mountain.

Back in France, Herzog was quickly turned into a national hero whilst the other men were sidelined. Under the terms of the contract no one but Herzog was allowed to write anything about Annapurna for five years. This was the first time that a 26,000-foot (8,000-metre) peak had been climbed and Herzog had a lot of personal charisma. Perhaps it wasn't surprising that he should get so much attention. In reality though, this was very much a team effort. Lachenal's self-sacrifice literally meant losing his toes for the sake of accompanying Herzog to the top. It wasn't his own ambition that prompted him to go on, it was his feeling of responsibility. Rebuffat and Terray gave up their own chance to reach the summit and risked their lives bringing Herzog and Lachenal down. They did this, not because they had sworn an oath of allegiance, but because of their own personal codes of honour, which demanded self-sacrifice if another climber needed help.

There are many Annapurnas in the lives of men.

MAURICE HERZOG

ANNAPURNA: CHRIS BONINGTON'S PARTY, 1970

It is interesting to compare this French expedition to Chris Bonington's attempt on the South Face of Annapurna in 1970. Technically, it was much more difficult than the route Herzog and Lachenal followed; at 9,750 feet (3,000 metres) the south face was one of the longest 'pitches' in the world. Bonington relished the challenge of a 'big wall' like this, but he realised that it would require a team that was prepared to work hard. He needed top climbers because it was such a technically difficult route, but he also needed them to do their share of 'plodding' because it was clearly going to be a long drawn out siege. Bonington chose carefully from amongst men whom he respected and had climbed with in the past. The only exception was a very talented American climber, Tom Frost, whom he took along to boost the expedition's international profile.

On the mountain, Bonington was careful to share the hard, but exciting work of leading, with the drudgery of carrying loads. He split the climbers into pairs and set up a system whereby each pair would spend some time at the front, then come down to base camp for a rest and then go up to do some load carrying between the camps. As they progressed higher and higher up the South Face, the climbers began to realise that the load carrying was as taxing as taking the lead, if not more so. It was absolutely necessary though because Annapurna had to be climbed in stages and that meant a continual re-supplying of the all the individual camps. Whereas on Herzog's expedition teamwork meant sharing the dangers, on this expedition it meant sharing the hard slog.

After several weeks Bonington started to worry that things were moving too slowly. The front pair was still far from the top and the monsoon was coming. Bonington himself was down at base camp, laid low by an attack of pleurisy. He realised that if they were going to make it this season, he would have to change the order of the climbers: up would go his best pair to leapfrog the men in front, whom he judged to be exhausted from load carrying. Naturally these men were not pleased, and one member of the team accused Bonington of destroying the spirit of the expedition. Bonington agreed to compromise and delay the change over, but he knew that he needed his best men out in front at this crucial stage in the climb. Eventually it was this pair, Dougal Haston and Don Whillans, who actually reached the top.

In 1950 Maurice Herzog had led from the front, but in 1970 Chris Bonington took a much more strategic role, continually assessing the strengths of his team in order to use them to best effect. There were no oaths of loyalty; instead Bonington tried his best to be fair and to maintain a good team spirit. At a certain point though, he took the decision to break an established pattern in order for the expedition as a whole to succeed. Teamwork in his book did not mean simple egalitarianism; he was a democratic leader but he could be decisive when he needed to be. Some of the climbers were happier than others to sacrifice their own ambitions, but in the end it was clear to everyone that the success of the climb was down to a genuine team effort.

DISCIPLINE

Good discipline is vital to effective teamwork. Expeditions are often very stressful environments and leaders can have to make difficult choices; they need to be able to rely on their team to follow their decisions. In modern expeditions, discipline has to be enforced collectively; in the past it was simpler and more brutal.

THE VOYAGE OF THE *ENDEAVOUR*, 1768-71
EXTRACTS FROM CAPTAIN COOK'S DIARY
* * *

November 19th 1768, Rio di Janeiro: Punished John Thurman, Seaman, with 12 lashes for refusing to assist the Sailmaker in repairing the Sails.

December 30th 1768, Rio di Janeiro: Punished Robert Anderson, Seaman, and William Judge, Marine, with 12 Lashes Each, the former for leaving his Duty ashore and attempting to desert from the Ship and the latter for using abusive language to the Officer of the Watch, and John Reading, Boatswain's Mate, with 12 lashes for not doing his duty in punishing the above men.

April 16th 1769, Tahiti: Punished Richard Hutchins, seaman, with 12 lashes for disobeying commands.

April 29th 1769, Tahiti: Punished Hy. Jeffs, Seaman with a dozen lashes for ill-behaviour on shore. He had been rude to a man's wife yesterday, of which the Indian complained, and Jeffs was confined immediately the Captain had the fact plainly proved, and next morning the captain had the offended parties on board, who were ignorant of his intentions. All hands being called, and the Prisoner being brought aft, the Captain explained the nature of his Crime in the most lively manner, and made a most Pathetick speech to the Ship's company during his punishment. *(Taken from the log of Mr Molineux, the ship's master.)*

June 4th 1769, Tahiti: Punished Archd.

Wolf with 2 dozen lashes for Theft, having broken into one of the Storerooms and stol'n from thence a large quantity of Spike Nails. *(Nails were used to buy sexual favours from Tahitian women.)*

June 12th 1769, Tahiti: Yesterday complaint was made to me by some of the Natives that John Thurman and James Nicholson, Seamen, had taken by force from them several bows and arrows and plaited hair, and the fact being proved upon them they were this day punished with 2 dozen lashes each.

June 19th 1769, Tahiti: Punished James Tunley with 12 lashes for taking Rum out of the Cask on the Quarter Deck.

July 14th 1769, en route to New Zealand: Punished the 2 Marines who attempted to desert from us at George's Island with 2 dozen lashes each and then released them from confinement.

November 13th 1769, leaving Mercury Bay, New Zealand: Samuel Jones, Seaman, having been confin'd since Saturday last for refusing to come upon deck when all hands were called, and afterwards refused to comply with the orders of the officers on deck, he was this morning punished with 12 lashes and remited back to confinement.

November 30th 1769, North Island: I order'd Matthew Cox, Henry Stevens and Eman Parreyra to be punished with a dozen lashes each for leaving their duty when ashore last night and digging up Potatoes out of one of the plantations. The first of the three I remitted back to confinement because he insisted that there was no harm in what he had done.

December 1st 1769, North Island: Punished Matthew Cox with 6 lashes and then dismiss'd him.

February 21st 1771: In the morning punished Thomas Rossiter with 12 lashes for getting Drunk, grossly assaulting the officer of the Watch, and beating some of the sick.

Expedition discipline in the eighteenth-century Royal Navy was harsh and uncompromising. Captain Cook's journal for his epic world voyage in the *Endeavour* reveals a world where ill discipline was punished by flogging and incarceration. Captain Cook knew that he had to keep a tight rein on the crew: this was set to be a long, potentially difficult voyage and it was important for him never to have authority questioned. It is notable though that after a spate of floggings at the beginning, between December 1769 and February 1771 there were none at all. Cook was a very unusual figure in the Royal Navy. He had worked his way up through the merchant marine before joining the military so he didn't have the brutal 'us and them' attitude of most other Royal Naval captains; he preferred to earn the respect of his men rather than flog them into submission.

Modern expeditions, thankfully, are not ruled by the threat of corporal punishment. Discipline has to be maintained by consensus rather than by fear; obviously this is far preferable but it doesn't lead to maximum efficiency. Ed Webster's account of his first Himalayan expedition shows how different modern ideas about expedition discipline are. In 1985 he was invited to become a member of an American team which was planning to make an attempt on Everest's notoriously difficult West Ridge. On the walk in to base camp Webster found himself being criticised by the other

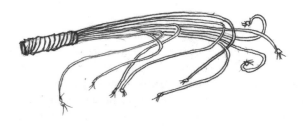

members at a team meeting for not pulling his weight. It was a fair cop; this was Webster's first big expedition and he admitted to not being used to all the endless packing and re-packing that had to be done on the way in to the mountain.

The team had two leaders: Robert Saas organised the expedition and Jim Bridwell was appointed as the climbing leader. Altogether, there were 21 members. On the mountain, Bridwell sometimes barked orders but a lot of decisions were made democratically or by simply letting things happen. As the expedition started up the West Ridge there were continual arguments, with various members of the team periodically accusing each other of not doing enough, or of presuming too much.

They were not a highly disciplined team and even the Sherpas re-marked that individual climbers seemed to go up and down the mountain almost at will; this was the American way, the team joked. The difficulty with this approach was that as the expedition progressed, the large team became more and more strung out on the mountain; the opportunities for team meetings therefore declined, so it was harder to enforce a spirit of collective responsibility. Ultimately none of the team members managed to reach the summit, but they did get further on the mountain than the last two American expeditions.

Webster later wrote that complete selflessness was one of the qualities needed by the members of an Everest expedition, but as he came to realise, this ideal was not so easy to achieve.

MORALE

A good laugh doesn't require any additional weight but counts for so much on any expedition.

ERNEST SHACKLETON

Any good leader knows that maintaining morale is crucial to the success of an expedition, especially when things are going badly. Though histori-cally the Royal Navy is often remembered for the harsh punishment meted out on errant sailors, some captains were quite sophisticated in their approach to maintaining crew morale, particularly on long expeditions to the Arctic and Antarctic. In one corner of the ship there might be a box containing several whips and a cat-o'-nine-tails, but in another you might find a stash of theatrical costumes.

The great tradition of naval amateur dramatics started on Captain William Parry's voyage in search of the Northwest Passage in 1819. As winter approached and his two ships were frozen in, he decided to put on plays to keep the crew amused, cajoling his officers into donning improvised fancy dress. He even took part himself, on the basis that it was his duty to set 'an example of cheerfulness'.

The experiment was deemed to have been such a success that on his second voyage in 1824 Parry added costumes and stage scenery to his ship's equipment. Thereafter theatrical entertainments became a standard part of British polar expeditions. They kept the crew and officers busy, broke the monotony of shipboard life and introduced a bit of fun into what could otherwise be very tedious voyages. In the summer when ships would break free of ice, there would be no time for theatricals but during the long winters they were very important for boosting morale. The tradition started to die out in the early twentieth century. On Captain Scott's first Antarctic expedition in 1901 the thespians on board took over a hut where they staged concerts, slide shows and a self-written play but by the time of his second expedition in 1910, the officers and scientists were keeping themselves amused at night by taking turns to deliver lectures on their chosen fields and personal interests.

Another feature of polar voyages was the expedition 'newspaper'. Again, the tradition started with Captain Parry. On his first expedition, he founded the *North Georgia Gazette and Winter Chronicle*, a handwritten weekly newspaper which dealt with, as Parry put it, subjects applicable to our situation. Some of his officers considered it witty enough to have reprinted when they returned to London. This practice was adopted by other polar explorers and when Ernest Shackleton went south in 1907 for his attempt on the Pole, he took a whole printing press with him.

EXPEDITION NEWSPAPERS

NORTH GEORGIA GAZETTE AND WINTER CHRONICLE
Edward Parry's two expeditions in search of the North West Passage

LITTLE AMERICA TIMES
Richard Byrd's expeditions to Antarctica 1933–1935

SOUTH POLAR TIMES
Robert Falcon Scott's expeditions to Antarctica in 1901 and 1910

AURORA AUSTRALIS
Ernest Shackleton's expedition to Antarctica in 1907–1909

FRAMSJAA NANSEN
Fridtjof Nansen's Polar Drift expedition 1903–1906

READING MATTERS

In 1902, in a remote village in northern Pakistan, the eccentric British climber Aleister Crowley had one of the strangest arguments in the history of exploration. Oscar Eckenstein, the co-leader of his expedition to the Karakoram Range, demanded that Crowley should abandon his large case of books in order to reduce their baggage. It seemed like a reasonable request as they were about to trek up the Baltoro glacier but Crowley refused point blank to do so and threatened to leave altogether; as he maintained later, he would 'rather bear physical starvation than intellectual starvation'. Eventually Crowley won the day.

He wasn't the only explorer to recognise the importance of books. Henry Morton Stanley took seven tonnes of them on his second expedition to Africa. Polar expeditions frequently took large libraries; Ernest Shackleton was fond of reading Robert Browning, and Robert Falcon Scott was partial to the works of Charles Darwin.

Peter Freuchen, the Danish explorer, once lost a whole crate of books at the beginning of a long hunting trip in Greenland. He was inconsolable until an Eskimo from a local settlement tracked him down with a book that had been recovered from the ice. For the next couple of months his bedtime companion was the exciting-sounding tome, *The Relationship between Denmark and the Popes at Avignon* by Dr L. Moltesen. Reading it night after night, Freuchen reached the stage where he knew most of the book by heart. Slowly he began to hate it and to fantasise about killing its author. Years later when he returned to Denmark, a dinner was given in his honour and to his amazement he found himself next to Dr Moltesen. After several hours, trying to keep his inner rage at bay, Freuchen introduced himself to the author and told him the story of the book, leaving out his homicidal fantasies. Subsequently Freuchen had his copy of the book lavishly rebound before presenting it to the author, sea water and oil stains included.

Living Together

If we could only get away from each other for a few hours at a time, we might learn to see a new side and take a fresh interest in our comrades; but this is not possible. The truth is that we are at this moment as tired of each other's company as we are of the cold monotony of the black night ... To sleep is our most difficult task, and to avoid work is the mission of everybody. Arctowski says, 'We are in a mad-house,' and our humour points that way.

THROUGH THE FIRST ANTARCTIC NIGHT,
FREDERICK COOK

One of the major stresses on polar expeditions is the frequent necessity to share cramped accommodation for a long time. When the pressure gets too great, this can lead to 'cabin fever', a morbid, irritable depression which often strikes in the winter.

Frederick Cook was the expedition doctor on board the *Belgica*, a Belgian ship which went to Antarctica in 1897 and spent over a year stuck in the ice. He wrote a fascinating account of how the crew was affected by their confinement. As the searing glare of the pack gave way to the stifling darkness of the polar winter, Cook recorded how the men became physically and emotionally depressed. They all began to suffer from what he called 'polar anaemia': they were pale and listless, uninterested in food or sleep. No one wanted to do any work and all attempts at raising their spirits failed. One sailor literally went mad and an officer died of a recurrent heart condition. The psychological pressure of their close-quarters existence wasn't eased until the sun came over the horizon and the men were able to climb off the ship and make excursions across the ice. Unlike the British captains who were always thinking of morale-boosting schemes when they had to over-winter, the *Belgica's* captain hadn't made any preparations and his attempts to break the monotony invariably fell flat.

A large part of the problem was that the captain hadn't told the crew in advance that he was planning to spend the winter in the Antarctic. If they had been given a proper chance to prepare and adjust themselves psychologically, their confinement would not have been so hard for them. It was very different for other explorers who had built into their schedules a period of over-wintering.

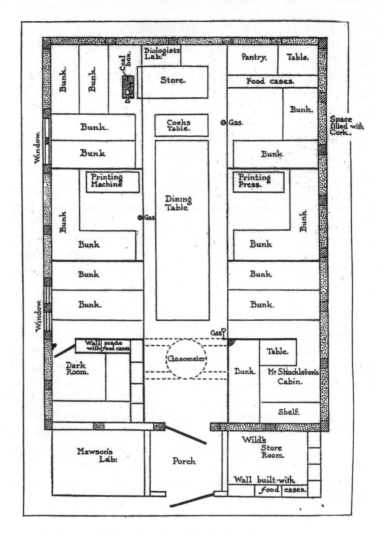

SHACKLETON'S WAY:
GOOD HOUSEKEEPING

In 1901, when Ernest Shackleton took part in Robert Falcon Scott's first Antarctic expedition, the officers and the crew kept to their own separate messes, but when Shackleton came back in 1908 as the leader of his own expedition, there was much less formality in the layout of the expedition hut. Shackleton had his own room, a 6 x 7-foot (1.8 x 2.1-metre) cubby hole, whilst the other 14 men paired up to live in cubicles separated by cloth walls. These cubicles were idiosyncratically decorated according to the tastes of the occupants. They were soon known by their nicknames: 'The Rogues' Retreat', 'The Pawn Shop', 'No.1 Park Lane'. Each man customised his own bed: some lay on piles of petrol tins, others on improvised hammocks. A curtain at the front gave the cubicles a little privacy allowing the men, as Shackleton put it, to 'sport their oak' if they needed (this wasn't quite as interesting as it sounds: it simply meant to be alone).

Shackleton may have had a separate room but it doubled as the expedition library, so his door was usually open. The men all ate together around a dining table made out of old packing cases which could be hoisted up to the ceiling between meals to maximise their space. They shared the hut chores and took turns at playing night watchman and mess-man. Shackleton recognised that there were dangers in a 'lack of occupation' so the small hut also housed a dark room, a laboratory and a printing press which had been donated to the expedition by a well-wisher.

Tensions did occasionally come to the surface. At the beginning of August, Mackay, one of the surgeons, attacked Roberts, the cook, for putting his boots on his sea chest. Fortunately there were other men on hand to keep the two apart and, fortunately too, the polar winter was coming to an end. As the men began to prepare for their summer sledging, they soon had plenty of others things to occupy themselves with.

ACHIEVEMENTS OF SHACKLETON'S
FIRST ANTARCTIC EXPEDITION
* * *

1. A new record for the furthest south: 88°23'S, 97 nautical miles from the South Pole

2. The attainment of the magnetic pole

3. The first ascent of Mt Erebus

By anyone's count, Shackleton's first expedition was very successful. In particular he won great public acclaim for getting to within 97 miles (156 kilometres) of the South Pole; his fellow explorers, Amundsen and Nansen, were equally impressed by Shackleton's decision to turn back when he realised that he and his men didn't have enough supplies to make a successful return trip. Shackleton desperately wanted to get to the Pole but he didn't think it was worth sacrificing anyone's life to get there. As he wrote to his wife, 'a live donkey is better than a dead lion'.

Shackleton's reputation as a great leader was sealed on his next expedition which was anything but successful. There are many stories of catastrophic expeditions to the polar regions which climaxed with amazing escapes from the ice, but none of them has had the continual power to fascinate as the epic story of Shackleton's second expedition. It is a wonderful adventure story which contains many lessons in leadership and teamwork.

Crisis Management: The Endurance

We had pierced the veneer of outside things. We had 'suffered, starved and triumphed, groveled down yet grasped at glory, grown bigger in the bigness of the whole'. We had seen God in all his splendours, heard the text that nature renders. We had reached the naked soul of man.

SOUTH, ERNEST SHACKLETON

In 1914 Shackleton went south once again, hoping to make what he called 'the last great polar journey' by crossing the Antarctic from coast to coast. After just a few months the *Endurance* became trapped in the ice; at first Shackleton wasn't too worried, expecting that eventually it would float free. After 281 days, he gave the order to abandon ship when the *Endurance*

started to buckle under pressure of the ice. The ship lasted for a few more weeks before being sucked down into the depths of the ocean.

Shackleton and his men were in a difficult, but not quite desperate, position. The nearest land, Paulet Island, was 346 miles (557 kilometres) away; Shackleton knew that a supply depot had been left there by an earlier expedition. The crew managed to salvage three small boats and quite a lot of supplies from the *Endurance,* before it sank. Shackleton hoped they would be able to walk to safety across the ice floes, but when they tried to they found it was backbreaking work. The surface of the ice was treacherous and inconsistent, so after barely a week Shackleton decided to call a halt and set up camp on an ice floe.

Shackleton's hardest task during this period was to maintain his men's morale. They nicknamed him 'The Boss' but he was far from autocratic. Shackleton could see that many of the men were very worried, but by sheer force of his personality, he tried to keep everyone optimistic. Frank Worsely, the captain of the *Endurance,* later wrote that when things were going well, Shackleton could be irritable, but he was never short-tempered when things were going badly. Shackleton felt that it was his absolute duty to bring everyone back alive and he tried to make everyone feel that he was the man that could do it. Shackleton was as tough as they come, but Worsely saw a 'touch of the woman' in the way he looked after his people. He was always quick to chat and share a joke; he took a personal interest in his team to the point of what Worsely described as 'fussiness'. The Boss was undoubtedly a 'man's man' but he had spent much of his life around women; he had eight sisters and several mistresses, as well as a dutiful wife.

As they languished on the ice, there were funny moments and there were difficult moments. At one point, Shackleton donned a full naval dress uniform and marched around the camp staging a mock inspection. On another occasion, he really did have to play it tough. When Tom McNeish, the carpenter, refused to carry out one of the officer's orders, Shackleton assembled all the men together and read them the Ship's Articles, the rule book for merchant seamen. Then he took McNeish to one side and warned him that if he had to, he would shoot him. Shackleton instinctively wanted everyone to follow orders because they respected him and the other officers, but he also knew that this was no place for divisions: if they didn't all stick together under Shackleton's leadership, no one was likely to survive.

After three months, the ice began to break up, and they had to abandon their camp and take their boats. The currents took them towards Elephant

Island, a desolate outcrop of rock and ice, whose beaches were permanently lashed by the sea. It was a remote, unforgiving spot which was unlikely to be visited by any other ships. Shackleton had to act decisively. Leaving most of the men behind, he took a party of six on a hair-raising voyage towards South Georgia, where there were several whaling stations. Frank Wild, his second-in-command, was left in charge of the men on Elephant Island; perhaps as a precaution though, Shackleton took McNeish, his truculent carpenter, with him.

During both boat journeys, Worsely again noted how Shackleton was always much more concerned for his men than for himself. He joked with them, rolled cigarettes for them and dished out hot drinks to anyone he thought was flagging. While others slept, Shackleton always tried to stay awake. And he was sensitive to nuances of individual behaviour; having been on two previous Antarctic expeditions, Shackleton knew that small irritations could easily turn into major fallings out, and that the way to stop this from happening was to demand that everyone should be thoughtful and considerate. He even told his men not to swear; at every moment, he insisted, they should be trying to keep each other cheerful. It wasn't simply a one-man band: Shackleton placed a lot of reliance on both Frank Worsely and Frank Wild, and both responded by raising themselves to the challenge. Shackleton's 'can-do' attitude brought the best out of people and his sheer hard work inspired them.

After 16 freezing, storm-lashed days, Shackleton and his men reached the island of South Georgia only to find that they had landed on the wrong side. So Shackleton split the party again and took Worsely and Tom Crean on an even more dangerous crossing of its uncharted mountains; they were equipped with home-made hobnail boots and a carpenter's adze for an ice-axe. Again Shackleton simply wouldn't give up; no matter what went wrong he always had a plan and always seemed confident that he would succeed. Somehow, through good fortune and sheer grit, they managed to make it to the other side and four months later Shackleton was finally able to go back and rescue the remainder of his men from Elephant Island.

A lot has been written in praise of Shackleton, though as an expedition organiser he clearly had his failings. He wasn't very good with money and, like Robert Falcon Scott, he initially underestimated the value of Eskimo travel methods in the polar regions. He was, however, a very effective man-manager. Crucially, he understood that it was as important to inspire people as it was to give them orders.

SHACKLETON'S CHRISTMAS LUNCH, 1915
★ ★ ★

Stale thin bannock
A mug of cocoa

ROBERT PEARY'S CHRISTMAS DINNER, 1891,
IN A HUT ON THE COAST OF GREENLAND
★ ★ ★

Whisky cocktail
Rabbit pie with green peas, corn and tomatoes
Venison with cranberry sauce
Plum pudding with brandy sauce
Apricot pie
Pears
Coffee with candies, nuts and raisins

HENRY MORTON STANLEY'S CHRISTMAS FESTIVITIES
DURING HIS SECOND AFRICAN EXPEDITION
★ ★ ★

Canoe races
Foot races between members of Stanley's team
Foot races by local women
A dance by 100 members of the Wanyamwezi tribe

When Teams Go Wrong

It might seem obvious that for an expedition to succeed it needs a good leader and a strong team, but it is remarkable how many expeditions have neither. There are many different reasons for poor teamwork, but there are some common patterns.

AD HOC TEAMS

Ad hoc teams are always a risk; if you are going to be successful in any co-operative venture then you really do need to know who you are working with. On mountaineering expeditions, where members are literally roped together, ad hoc

teams are particularly dangerous. The story of the first ascent of the Matterhorn illustrates this with appalling clarity.

The Matterhorn is a very striking mountain on the border of Italy and Switzerland. During the early 1850s and 1860s, many of the peaks in the Alps were climbed by British mountaineers but no one could find a way up the Matterhorn. Edward Whymper, a young engraver from London, became obsessed with it; he was determined to be the first man to reach its summit and made numerous attempts. In 1865 he went on a very successful tour of the Alps, making first ascents of several unclimbed peaks before heading for the Matterhorn for what he hoped would be the last time. He arranged to make another attempt from the Italian side with a well-known Italian guide, Jean-Antoine Carrel; Whymper had climbed with Carrel many times and knew him well. What he didn't realise was that Jean-Antoine had ambitions to climb the Matterhorn with an all-Italian party. On the morning they were due to begin, whilst Whymper slept in his bed, Carrel set off without him.

Whymper was livid but there was nothing he could do. There wasn't even a single porter to carry his bags. Then, by chance, he met another British climber, Lord Francis Douglas, who offered to help him take his ropes and climbing equipment over to the other side of the Matterhorn in Switzerland. On the way, they concocted a plan to make an attempt on the Matterhorn together, from the Swiss side. Then, when they reached Zermatt, they encountered another well-known British climber, Charles Hudson; he was touring the Alps with one of his former pupils, Douglas Hadow. Whymper discovered that Hudson too was planning to make an attempt on the Matterhorn, so he felt that the best thing to do was to join forces with him and make one big party. Hudson agreed, on condition that his young protégé should come along too. Whymper wasn't keen on taking Hadow; neither he, nor Lord Francis Douglas for that matter, had very much experience and this was going to be a particularly difficult ascent because so little was known about the Swiss side of the Matterhorn. But Whymper was over a barrel: his Italian rival Carrel was already on the mountain, and Hudson and Douglas had already booked all the best local mountain guides. If he didn't accompany them, then someone would beat him to his prized summit. And so Edward Whymper, who was usually as careful in his climbing as he was in his work as an engraver, found himself heading up the Matterhorn with three climbers that he had never met before on a mission to climb a mountain that many people thought was impossible.

To everyone's great surprise, Whymper's party discovered that in fact, there was a comparatively easy route to the top on the Swiss side. They spent almost an hour on the summit, long enough to spot the Italian party below and throw stones at them. Whymper felt triumphant: on his eighth attempt he had conquered the Matterhorn and he had beaten Carrel to it. On the way down, everything changed. Hadow slipped and dragged Michel Croz, Charles Hudson and Lord Francis Douglas after him. A few moments later they were smashed to pieces on the glacier below; one body was never found. Whymper survived, but he was psychologically scarred for the rest of his life. The whole episode showed how easy it was for accidents to happen on the way down but it was also a classic case of an ad hoc team coming together, surprising themselves at how well they get on at first and then falling apart at a later stage.

NEW TEAM MEMBERS

There is always danger when changing a team, especially if some of the members are very set in their ways. This is particularly important in small groups.

Peter Freuchen came across a case in Greenland of a fur trapper, Gustav, who fell out with his long-standing partner and took on a new partner, Olav, for the winter. The two men went to a remote point on the coast of Greenland where the trapper had a hut. At first it all went well: Olav was young and enthusiastic and much more talkative than the trapper's former partner. After a few months though, he fell ill and died. So Gustav excavated a shallow grave and buried Olav under some stones to keep the wolves away.

Gustav knew that it would be almost another year before the ship was due to collect him and his cargo of furs; as the nights drew in and the Arctic winter arrived, he became terribly lonely, so lonely that he dug up Olav's corpse and took it back to the cabin where he sat it down at the kitchen table. Overcome by shame, he reburied Olav's body the following morning, only to dig it up again a few days later. When the summer came, the corpse began to thaw out and stink, and Gustav realised that he couldn't keep it in the cabin. By now though, he was being tormented by visions of the dead man's ghost. So, to put an end to him completely, Gustav took out his rifle and shot the dead man in the head. Then he buried him again.

When the ship finally came in, the corpse was exhumed and the trapper taken back to Denmark, where he was accused of murder. An autopsy revealed that, as Gustav maintained, his partner was long dead before he shot him. However, by the time the charges were dropped, Gustav had been declared insane and placed in an asylum.

Freuchen put the events down to Gustav's inability to deal with a change of routine; he had been with his previous partner for many years and had got used to his taciturn ways. When Olav came along, life was more interesting and Gustav talked like he never had before. So when

the man died, it was a huge blow which ultimately sent Gustav over the edge into madness. If his new partner had been more taciturn, the trapper might not have been so unsettled by his death.

COMPETITIVE TEAMS – ROWELL

Competitiveness is the antithesis of team spirit but it is often a factor in large groups. Galen Rowell's account of an American K2 expedition in the 1970s, In the Throne Room of the Mountain Gods, *gave a candid insight into the tensions which can undermine large expeditions.*

By the mid-1960s, all of the major Himalayan peaks had been climbed and a new generation of mountaineers had emerged. They were more willing to take risks, and more individualistic. The story of the American expedition to K2 in 1975 shows how destructive that individualism can be.

The expedition was put together by Jim Wickwire, a lawyer from Seattle. He invited another Jim, Jim Whittacker, to be the leader. Whittacker was famous as the first American to climb Everest but he had never before led an expedition on this scale and most of his mountaineering had been done in the US. Wickwire and Whittacker recruited six other climbers, two stills photographers and a film cameraman. Apart from Whittacker, none of them had been to the Himalayas before; in theory they were chosen for their compatibility but, as a sign of things to come, one of the original members of the team was ignominiously dropped before they left America.

As the long trek in to the foot of the mountain got underway, the tensions within the team quickly came to the surface. At this very early stage, they started to speculate on who would take part in the summit attempts. Usually this would not be addressed until much later, when an expedition actually started to climb properly, but several members of this team saw things differently. They even began to compete with each other over who could carry the heaviest back packs. Galen Rowell, one of the expedition photographers, usually carried a light rucksack in order to be able to run ahead of the group if he saw any good photo opportunities. To his disgust, there were soon comments flying about on how little he could carry.

By the time they reached base camp, the team had fragmented into cliques. Several of the climbers were ill with bronchial and stomach disorders and tempers were fraying. 'Big' Lou Whittacker, the brother of the expedition leader, threatened to 'deck' 5 foot 8 inch (1.7 metre) Fred Dunham after a trivial misunderstanding about taking down a tent. Dunham fantasised about replying with an ice-axe, to the back of Big Lou's head.

Finally they began to make forays on to the unclimbed North West ridge; earlier, they had managed to make a brief aerial reconnaissance of K2 and it looked like a feasible route. The bickering didn't stop, though, when they got to work, and neither did the competition. The most powerful clique was nicknamed the 'Big Three'. It seemed to the others that the Big Three (Jim Whittacker, Lou Whittacker and Jim Wickwire) always wanted to be in the lead and that they presumed they were going all the way to the top. Eventually the tension came to the surface with a full-scale confrontation; two members of the team, Fred Dunham and Fred Stanley, threatened to leave because they felt so alienated.

The 'Two Freds' stayed put, but over the next few weeks the schism between the Big Three and the others refused to heal. Constant bad weather added to the gloomy mood and low morale. In spite of this, the team continued to try to find a way up the ridge, with the Big Three almost always in the lead. It was becoming more and more obvious, though, that this ridge wasn't going to take them all the way to the summit. Eventually they were forced to admit defeat and return to America, licking their wounds. In a bizarre final twist, a rumour began to circulate that the expedition was, in fact, funded by the CIA; the story went that Whittacker's team had been asked to plant a top-secret listening device on the Chinese border. For once they all came together to reject this charge as absurd. As Galen Rowell later wrote:

Now that there were no laurels to share we realised that we had far more in common than we had presumed. No matter how badly our team had failed on the mountain or in personal relationships, a core of dignity remained. That dignity was not rooted in what we had done but what we understood about ourselves. When the cloak of potential achievement was removed, one thing remained: a desire to climb the mountain.

The rumour was quickly discredited but the realisation of their common core of dignity came far, far too late. The 1975 expedition went down in mountaineering history as one of the most fractious ever mounted, partly because the members agreed to hand over their personal diaries to Galen Rowell. In those days, most expedition books papered over any disagreements, but *In the Throne Room of the Mountain Gods* was brutally honest. Ultimately the expedition was defeated not because of poor teamwork, but because they chose the wrong route. It was, though, an unpleasant experience for almost everyone, leaving them with bitter memories. Remarkably, three years later Whittacker and Wickwire went back to K2 with two other members of the 1975 expedition to make a second attempt. Clearly their desire to climb the mountain *was* greater than any mutual antipathy, and this time they succeeded.

Mutiny

It is every leader's worst nightmare – the point when their team unites, to turn against them.

In 1909 Claude Rusk led a party from the Mazama mountaineering club of Portland on an expedition into the wilds of Alaska. His aim was to go over the controversial route of Frederick Cook, an explorer from New York who claimed to have climbed Mt McKinley, the highest peak in America. The other men in his party were intimidated by the landscape and it wasn't long before they wanted to get back home. Rusk would hear none of it, so the others concocted a plan to kill him: they hit him over the head with a snowshoe and pushed him into a crevasse. It was a fairly half-hearted attempt though and Rusk survived; he even forgave his would-be assassins. They carried on the search and when Rusk returned to the Portland, he told no one but his wife.

In 1871 Charles Francis Hall, an American explorer, set off for the North Pole. He was a patriot and self-taught man who believed that it was his destiny to reach the North Pole and claim it for the United States. Hall was persuaded to take two German scientists on his ship, the *Polaris*, but from the very beginning they were disrespectful towards him. The ship's captain was an alcoholic; he was a friend of Hall's but he was very nervous about the whole voyage.

When the *Polaris* reached the pack ice, Hall made a preliminary sledging trip northwards to assess the ice. He returned to the ship full of confidence and enthusiasm but after being offered a cup of coffee, Hall suddenly fell ill. When he recovered briefly a few hours later, he accused one of the Germans, Dr Emil Bessels, of having poisoned him. After that Hall refused to eat any food which had not been prepared by two faithful Eskimos that he had employed to assist him. Within a few weeks though, he was dead. There were some members of the crew who believed that Hall was the victim of a conspiracy but his personal diary mysteriously disappeared and in spite of an official enquiry, nothing was ever proven. In 1968 the American writer Chauncey Loomis tried to resolve the mystery by going to Greenland to exhume Hall's body. He left with samples of Hall's nails and hair, which were submitted for laboratory analysis in Canada. They were subsequently found to contain toxic levels of arsenic, which the scientists calculated to have been consumed within the last two weeks of Hall's life.

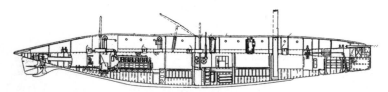

THE SUBMARINE "NAUTILUS"

In 1931 Sir Hubert Wilkins bought a First World War submarine and attempted to take it under the Arctic ice on a voyage to the North Pole. Wilkins was a visionary explorer who had previously made the first trans-Arctic flight; he christened the submarine the *Nautilus* in honour of Jules Verne. Wilkins had no trouble hiring a crew, but as they sailed across the Atlantic, the submarine continuously broke down. Several men began to have second thoughts. On their final stop at Bergen in Norway, someone swam under the submarine and sawed off its diving rudders. This act of benign sabotage was the equivalent of cutting the steering wheel off a car. Sir Hubert didn't find out until the *Nautilus* reached the Arctic, but he was characteristically undaunted and insisted on carrying on, even though the missing rudders made it virtually impossible for the expedition to proceed. Sailing on the surface, the *Nautilus* set a new record for the furthest north that any vessel had ever been; then Wilkins prepared to dive. In the event it was an anticlimax, albeit a frightening one. The *Nautilus* head-butted some ice floes and then managed to partially submerge, only for a large tungsten drill on top of the submarine to get stuck, underneath the ice floe. In theory this drill was to be used to cut air holes, but like a lot of other things on the *Nautilus*, it wasn't working properly. Sir Hubert shot some underwater footage, the submarine wriggled free of the ice and then they turned back.

*   *   *

A common thread running through many mutiny stories is the presence of a leader who is more willing to take risks than the rest of his team. The simple lesson is that prevention is better than cure. If you select your team well and make sure that they know what they are getting into, then mutinies rarely occur, unless you are grossly incompetent. If you ask people to take risks which they neither understand nor feel comfortable with, then you are asking for trouble.

Solitude

Whilst the pressures of working as a team may create problems, exploration can also involve long periods of solitude which are equally demanding.

In 1931 **August Courthauld** became the focus of an international rescue mission when newspapers reported that he had become trapped beneath several tonnes of snow on the Greenland ice cap. Courthauld was part of a British team which had gone to Greenland to examine its potential as a stop-over point for transatlantic flights. The team's leader, Gino Watkins, believed that the quickest way to get from Europe to the West Coast of America was to fly across the Arctic. One element of the expedition was a weather station which was set up in the middle of the Greenland ice cap. This was intended to house two men, but it was always difficult to keep properly supplied. In December of the previous year, August Courthauld volunteered for a stint as its sole occupant. He expected to be there for two to three months, but he had to stay for almost six. His last few weeks were spent incarcerated under the snow, when huge drifts covered the entrance

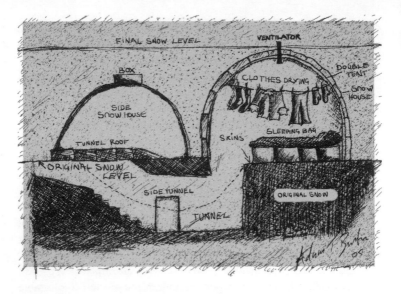

to the weather station and piled up above its roof. When his rescuers eventually arrived, the only visible sign of a human presence was a small ventilator pipe, poking up above the snow.

Before he became snowed in, Courthauld was kept busy looking after the meteorological instruments. He had neither a radio nor a gramophone but he wasn't bothered. At first he found the silence outside so profound that he didn't like breaking it; later he came to enjoy 'a sort of singing'. Interestingly Courthauld claimed that the longer his ordeal lasted, the more confident he was that it would end well, because he distinctly felt that 'some outer Force was on my side'. Looking back at the experience, Courthauld maintained that he couldn't really see what all the fuss was about and added that as long as it was done voluntarily, there was no reason why someone else shouldn't repeat the experience, in similar circumstances. Courthauld was a classic stiff-upper-lipped Brit, quiet and instinctively reserved. In a way he was the perfect man for such an experience. As he dryly noted, it wouldn't have suited someone 'of a nervous disposition'.

Richard Byrd, the American aviator, was very different. He was good-looking, gregarious and loved the limelight. Yet, just a few years after Courthauld, he too volunteered to spend a winter alone in a polar weather station; this time it was in the Antarctic, 123 miles (198 kilometres) into the

Great Ice Barrier. Courthauld's had been an impromptu decision but Byrd had time to plan. Again his main task was to collect weather data, but Byrd looked for-ward to his solitude as an opportunity to read a lot of books and listen to music. Unlike Courthauld he had a radio and was in regular contact with his expedition base, a collection of semi-permanent huts on the edge of the Ross Ice Barrier which was christened 'Little America'. As the leader of the expedition, Byrd was very concerned about being separated from the 55 men he had brought down to the Antarctic. His constant worry was that if anything went wrong, they might risk their lives trying to rescue him. When a build-up of carbon monoxide inside his hut poisoned him, Byrd pretended over the radio that nothing had happened. He could hear their voices, but all his talking was done in Morse code.

If at first his only battle was to stave off loneliness, Byrd's last months were spent desperately trying to stay alive. The kitchen stove and the generator for his radio both began leaking toxic fumes; Byrd found that he could just about bear the intense cold, but he had to eat and he had to keep in touch with his base. When at the end of the winter three members of his team finally arrived to relieve him, he didn't tell them what he had been through because he felt a deep sense of shame and didn't want to admit that he had been rescued.

Byrd gained an important leadership lesson from his self-enforced isolation: he discovered that he wasn't indispensable. The other members of his team got on very well without him and the fact was that, whether he liked to admit it or not, he *did* have to be rescued by them. From then on, he found it much easier to take a back-seat role in several aspects of his work. Like many people who have lived through traumatic experiences he returned home with a greater appreciation of the simple pleasures of being alive.

Invisible Friends

The experiences of August Courthauld and Richard Byrd demonstrate the huge resilience of, dare I say it, the human spirit. Solitude may be uncomfortable for long periods of time but, with the right attitude, it can be coped with. Interestingly though, such is the human desire to have company that when the chips were down, many mountaineers have found themselves helped by mysterious strangers.

Frank Smythe, Everest 1933

In June 1933, on the way down from an attempt on the summit of Everest, the British climber Frank Smythe had not one but two extraordinary visions. He was alone having just reached a high point of 28,200 feet (8,400 metres), equalling the world record for altitude. As he was descending, he sensed the presence of an unknown companion; Smythe felt as if he was connected to him by a rope and at one point even turned around to offer his invisible friend a piece of Kendal mint cake. Eventually his companion disappeared, but just before Smythe reached camp he saw two even stranger figures, hovering in the sky. One had wings, the other a kind of bulbous beak. Smythe stopped to test himself: he looked away but when he returned they were still there. He checked his alertness by identifying the peaks and valleys around him. Then as he moved off again, they disappeared into the mist. When he finally made it down to his companions, they were sceptical. The expedition doctor, Graham Greene's brother Raymond, described the strange apparitions as 'Frank's pulsating tea-pots'.

Stephen Venables, Everest 1988

On the way up to the summit of Everest the British mountaineer Stephen Venables was joined by a mysterious old man. Unlike Frank Smythe's imaginary companion, the old man was not there simply to look after his adopted mountaineer. Sometimes he needed support and Venables felt that he had to take care of him. When after reaching the top, he was forced to stop and spend a night out in the cold, Venables found himself being visited by several more imaginary friends. The great Eric Shipton warmed his hands whilst a whole gang of people looked after his feet. The old man stayed with him for the rest of the night and then disappeared at dawn. Venables gingerly made his way down and a few hours later he was shaking hands with two other members of his team.

Both Frank Smythe and Stephen Venables were climbing without oxygen so it may very well be that their invisible friends were hallucinations brought on by hypoxia. Smythe later wondered if his 'pulsating tea-pots' were a type of high-altitude mirage, created by light refracting through the air. Again, these experiences show how deep the human need for company is. Both Venables and Smythe had continued on when their real partners and teammates had been unable to keep going; their imaginary companions made up for the lack.

Meeting Natives

Knowing how to deal with strangers is as important to an explorer as knowing how to deal with the other members of his or her team. Apart from the Antarctic, there are very few regions of the world which are completely uninhabited. Local people can help expeditions in many ways: they can act as guides and porters, they can sell them food and supplies, they offer advice and directions and can teach outsiders how best to operate in a particular area. They can also cause a lot of problems by obstructing an expedition, and in the worst cases by attacking its members.

Most nineteenth- and twentieth-century explorers were interested in the indigenous peoples that they encountered on their expeditions. Though few of them had any training in anthropology, they were usually keen to record local habits and customs. Some, like the British explorer Richard Burton, recorded their encounters with academic detail; others

EUROPEAN NAMES FOR NATIVES
✱ ✱ ✱

Eskimo There are two possible roots for the word 'Eskimo'. For many years it was believed to be a Native American word meaning 'eaters of raw flesh', which was subsequently adopted by French–Canadian trappers. Today this pejorative etymology is in dispute, some linguists claiming that it actually derives from a word used to describe the way in which the Inuit laced their snowshoes. Modern Canadians and Greenlanders prefer to use the word Inuit to describe themselves; it is the plural of 'Inuk', meaning human being.

Indian This used to be the generic term for the natives of North and South America and is still widely used. Its use is the result of a geographical error by Christopher Columbus: when he arrived in Central America, initially he thought that he had reached India, so he referred to the natives as Indians.

Aboriginal It was used in Australia to distinguish white colonials born in Australia, who were called 'natives', from the real natives who occupied the land before the white settlers arrived. They were called Aborigines.

Coolie From Hindi 'quli' meaning hired servant. Coolie was used all over Asia to refer to a poorly paid servant or porter.

were content simply to describe the local colour in order to spice up their expedition narratives. As many explorers discovered, dealing with local people is not always easy but if they are treated properly it can be very rewarding. There are three groups in particular which played a crucial role in the history of exploration.

THE POLAR ESKIMOS

I been England, long time none very well. Long time none very well. Very bad weather. I know very well, very bad cough. I very sorry, very bad weather, dreadful. Country very difference. Another day cold, another day wet. I miserable.

LETTER FROM KALLI, A 17-YEAR-OLD ESKIMO TO A FRIEND; KALLI WAS TAKEN TO BRITAIN ON BOARD HMS ASSISTANCE. LIKE MANY VISITORS, HE WAS NOT IMPRESSED WITH THE WEATHER.

In 1818 the Scottish sea captain John Ross made the first recorded contact with the polar Eskimos on an expedition to the Arctic. Though southern Greenland had been visited by Danish and German missionaries for many years, very little was known about the far north of the island and the people who lived there. Ross's mission was to discover the Northwest Passage but he was keenly interested in meeting any indigenous peoples along the way. His ship carried boxes of presents including knives, bottles of gin, boxes of snuff and 40 umbrellas. On board there was also a native of south Greenland, John Sacheuse, who could speak Eskimo. He had learnt his rudimentary English on-board a whaling ship.

Almost four months into their voyage, the crew of Ross's ship spotted some figures in the distance across the ice. Ross sent out a small boat full of presents, but the Eskimos paid no attention and soon hurried away on their sledges. On the following day they came back. This time John Sacheuse was able to approach them tentatively; the Eskimos were very wary though. They told Sacheuse that at first they thought the British ships were enormous birds and they asked him whether he came from the sun or the moon. Eventually they agreed to come on board, and over the next few days several other Eskimos joined them to take a look. The 'Arctic Highlanders', as Ross called them, quickly overcame their nervousness and were happy to swarm over the ship, 'impertinently' examining all the strange objects they found, and occasionally attempting to pilfer them.

The men that Ross encountered were from a tribe that lived nearby on the Greenland coast; they became known to subsequent explorers as the Polar Eskimos. Although there were tribes scattered all over the Arctic, the Polar Eskimos had had no contact with anyone else for a long time. After a few days, Ross hoisted the flag and took possession of the region for the British Crown; he wondered if it might be possible to establish a trading relationship with the Eskimos, offering them iron in return for furs and ivory, but nothing came of it. Ross was fascinated by the Eskimos, but he seemed to regard them more as curios, than anything else.

During the first half of the nineteenth century, British explorers were very active in the Arctic and they encountered Eskimos on many occasions. In general though, they tended not to adopt Eskimo techniques. When George Nares led a British attempt on the North Pole in 1875, his sailors were equipped with heavy sledges which they were expected to drag across the ice. Nares had a local guide with him but he dismissed

Eskimo huskies as 'picturesque' and, instead of hunting for local food, he fed his men on standard naval rations. Needless to say, his expedition got nowhere near the Pole.

In the second half of the century, a wave of American explorers went to the Arctic. On paper, they were as condescending to the Eskimos as the British were, but in practice they behaved very differently. Men like Elisha Kent Kane, Charles Hall and Isaac Hayes recognised that the best way to deal with the Arctic was to copy the Eskimos even if it meant learning how to eat raw meat, mastering the art of dog sledging and wearing fur clothing. Robert Peary treated the Polar Eskimos almost as if they were his own personal property. In one of his books, he posed the rhetorical question: what is the purpose of the Eskimos? His answer was simple, and egotistical: he believed that it was his manifest destiny to reach the North Pole and that the Polar Eskimos were destined to be his assistants, or 'instruments' as he put it.

Peary made a huge impact on the lives of the Polar Eskimos; he was both respected and feared. 'Piulersuaq', as the Eskimos called him, took away their traditional source of metal, three huge meteorites at Cape York in Greenland, but he gave them boats, guns, knives and telescopes. He was jealous of *his* people, and ordered the Eskimos not to co-operate with anyone else who came to the Arctic. Peary's final expedition to the North Pole in 1909 involved 49 Eskimos. The men travelled with Peary's team, whilst the women played a vital role as seamstresses, for both their husbands and the white members of Peary's team.

It wasn't all just a question of exploitation though. Before the explorers arrived, the Polar Eskimos constantly teetered on the brink of disaster. When the hunting was good they thrived, but when it was bad, they starved. Infanticide was common and in times of famine whole families had been known to die of hunger. The American Elisha Kent Kane carried out the first census of the Polar Eskimos in 1856 and counted 140 men, women and children; he predicted that the tribe would soon be extinct. Perhaps that was why they seemed to be so compliant. Like many poor people, the Eskimos were supremely pragmatic and they invariably welcomed explorers into their midst. Sometimes it meant more mouths to feed, but usually the outsiders brought guns and knives, and that meant better hunting.

Though European and American explorers often presented themselves as benefactors, they invariably made sure that the Eskimos knew who was boss. In Elisha Kent Kane's famous account of his Arctic explorations, he reflected on how firearms maintained the respect of the Eskimos:

The power of a gun with a savage is in his notion of its infallibility. You may spare bloodshed by killing a dog or even wounding him; but in no event should you throw away your bullet. It is neither politic nor humane.

In other words, make sure that the Eskimos realise the power of your guns and don't ever let them think that you might miss. Even Roald Amundsen who respected the Eskimos hugely, occasionally felt it necessary to show off what he billed as the 'white man's superiority'. Whilst on his epic voyage through the Northwest Passage in 1903 he and his ship's lieutenant contrived a plan to demonstrate their power to the local Eskimos, in case the natives should ever turn against them. They laid an explosive mine some distance from their ship in an igloo and after gathering everyone together and lecturing them on the white-man's amazing powers, they blew it up. As Amundsen dryly noted: "This was all that was needed".

Ultimately, although the Eskimos were affected by their contact with European and American explorers, their lives were not really transformed until the 1950s and 1960s when the Cold War led to the militarisation of the Arctic. The construction of the Dew-Line radar stations and the huge American air base at Thule in Greenland brought more men and more money than ever before into the Arctic and this had a much greater impact than all the Pearys and Rosses of this world. The governments of Canada, Denmark and the Soviet Union all, to a greater or lesser degree, attempted to extend the boundaries of civilisation in the Arctic. The lives of today's Eskimos would be unrecognisable to nineteenth-century explorers.

THE SHERPAS

Like the Polar Eskimos, the Sherpas are a small ethnic group who became famous around the world through their contact with explorers and mountaineers. Their unique talent was, and still is, their ability to work well at high altitudes; this was exploited by successive teams of British mountaineers in their attempts to climb Everest. As a fitting climax to this relationship, Tenzing Norgay was one of the two climbers to reach the summit when in 1953 a British team made the first ascent. Today Sherpas are a still very much a part of Himalayan mountaineering and they

hold world records for the fastest attempt and the greatest number of ascents of Everest.

The word Sherpa means 'people from the east'; originally they were Tibetan farmers who had migrated to Nepal in search of better land. When Sherpas first encountered British mountaineers at the beginning of the twentieth century, they were poor but they were above the subsistence level of the Eskimos. The Sherpas were already a mobile work force; many of them had left their homeland in Nepal to hire themselves out as general porters or rickshaw drivers to the Indian merchants of Darjeeling. Alexander Kellas, a Scottish explorer and mountaineer, was the first to recognise the talents of the Sherpas as high-altitude porters. He took a group of them from Darjeeling on a climbing expedition to Sikkim and found they were more useful to him than the Swiss guides that he had brought all the way from Europe, at considerable expense. Kellas sang the Sherpas' praises and by the 1920s a contingent of Sherpas was a standard part of most Himalayan expeditions.

In Nepal, the Sherpas already had a reputation for being shrewd businessmen and they were quick to see the financial advantage of creating a niche for themselves as high-altitude porters. Local men might be employed to carry all the stores to the bottom of the mountain but Sherpas were paid more and kept on for much longer. Once they had established their credentials, Sherpas worked hard to keep the work within their small tight-knit community, resisting the inroads of other Tibetans and Nepalis. It was dangerous work, but it brought them prestige as well as hard cash; they didn't quite understand why European climbers were so willing to spend so much time and money on climbing mountains, but it didn't matter as long as they kept on coming back.

The climbers who most enjoyed working with Sherpas were those like Eric Shipton, who got to know them most intimately. He preferred to keep his expeditions small and there were certain Sherpas who worked for him on several occasions. At the end of one expedition, he took two of his favourite men, Ang Tharkay and Sen Tenzing, on holiday to India with his girlfriend. For Shipton, the Sherpas were perfect companions: hard working, smart and equipped with a good sense of humour.

Tenzing Norgay regarded himself as the first Sherpa who really wanted to climb for its own sake. He first came to prominence on a Swiss expedition to Everest in 1952. The Swiss warmed to Tenzing immediately and quickly recognised his talents as a mountaineer; they invited him to join the climbing team itself and not just work as the head Sherpa. Tenzing relished his new role and found the Swiss more affable and less hierarchical than the British.

Tenzing was exceptional though. It is worth remembering that although many of today's Sherpas are very technically advanced climbers, in the 1950s only a few of them had any of these skills. Yes, they had guts, determination and stamina but the British and Swiss climbers had to look after them on the mountain. Wilfred Noyce, one of the members of the British 1953 expedition, wrote afterwards that he found it very stressful to climb with Sherpas because he always had to be on his guard. This all began to change in the mid-1950s when the Himalayan Mountain Institute was set up in Darjeeling. Tenzing was the director of field training and several other Sherpas worked as tutors.

Perhaps inevitably as the fame of the Sherpas grew, they came to see Everest in particular as their patch. The first ascent in 1953 was followed by ugly arguments over who actually reached the summit first: Tenzing or Hillary? (*see* chapter five.) In 1978 when Reinhold Messner and Peter Habler climbed Everest without oxygen, a number of Sherpas were quick to denounce them as cheats, claiming that they had actually been carrying pocket-size cylinders. It was as if they couldn't believe that two European climbers could do what no Sherpas had achieved. Only when another team, which included some Sherpas, repeated the feat, did they accept that it was possible.

THE BEDOUIN

Bedouin comes from the Arab word 'bedu' meaning dwellers of the desert. There are many different Bedouin tribes: most of them live in north-east Africa and the Middle East. Today a lot of Bedouin live in towns and settlements, but for centuries they were a nomadic people who moved around the desert searching for water and pasture for their animals.

Unlike the Eskimos or the Sherpas, Arabs were frequently hostile towards Western explorers. They saw Europeans as unwelcome intruders, 'Ferenghi' descended from the Christians who had been their traditional enemies since the time of the crusades. When Richard Burton and Johan Burkhardt made visits to Mecca in disguise, they were breaking an ancient taboo against non-believers setting foot in the holiest city in Islam; they would probably have been killed if they had been found out. The Bedouin were fiercely Muslim but they considered themselves above the town Arabs, and had a slightly more tolerant position towards outsiders, provided they paid them due respect.

The Europeans who did attempt to explore the deserts of the Middle East were much more dependent on local guides than explorers of the Polar Regions or the Himalayan mountains. In many ways the term exploration was a misnomer: though the deserts might be blanks on European maps, they had invariably been explored by previous generations of Bedouin. For the British explorers Wilfred Thesiger and Freya

Stark though, this was part of the fascination; they were both huge admirers of the Bedouin and their journeys were infinitely enriched by the presence of Bedouin guides.

Thesiger was born in Ethiopia, the son of a minister at the British Legation in Addis Ababa. He went to school and university in Britain but after that he spent most of his time in Africa and the Middle East. Between 1946 and 1948 he made two crossings of the infamous Empty Quarter, a huge desert in the southern half of Arabia. Officially he was working for a locust control project but in reality these journeys were made with different motives. His Bedouin guides were intrigued; they assumed that he was really working for a foreign government, mapping their lands for future conquest. In reality it was all much more personal: these were inner journeys as well as geographical expeditions.

Thesiger had a very romantic view of the Bedouin, admiring their Spartan lifestyle and their strict codes of honour. According to his philosophy the harsher the environment, the better the people; Thesiger didn't just admire the Bedouin, he wanted to be like them. Throughout his desert journeys you get a sense of him trying to prove himself to the local people. When travelling with Bedouin guides, he insisted on eating the same food as they did and drinking the same water, or what little of it there was. While he had high regard for Bedouin traditions of hospitality, sometimes he found his guides exasperating. On one of his forays into the Empty Quarter, he went ahead with one of his Bedouin guides to a waterhole, because the other men could see that Thesiger was very thirsty. When they got there, his companion refused to drink anything in advance of the main party, and Thesiger didn't want to be the first person to quench his thirst. So, they sat there in the burning sun for two hours, waiting for the others to arrive. On another occasion, he and his guides caught a wild hare at the end of a long, journey across the Sands. They were all starving and looking forward to eating it when another group of Bedouin arrived in their camp. To Thesiger's amazement, his companions gave away *all* of the meat to their guests.

For Freya Stark, Bedouin hospitality could be equally exasperating. As she noted whilst travelling with them, it was impossible to know how much food to carry because even when crossing a seemingly empty desert there were always extra people turning up for dinner. Stark understood, though, the symbolic importance of food and she quickly realised that one of the best ways to make the Bedouin more sympathetic was to eat with them. On one occasion Stark was chased out of a small town by unfriendly

locals, and a gang of them followed her a long way into the desert. She told her guides to distribute food to the townsmen and their hostility evaporated.

Her perspective on the Bedouin and the town Arabs was rather different from Thesiger's. Freya Stark was much more interested in the social life of the people she met. As a woman she was allowed access to the harems as well as contact with the men. As she tended to travel much more publicly, Stark was required to fulfil a lot of social engagements. Clearly she enjoyed this, but on some journeys it became more of a duty than a pleasure. She particularly resented having to alter her route or make social calls at the end of the day, a responsibility that she regarded as being 'the chief hardship of travel'.

Both Stark and Thesiger realised that they were sometimes exploited by their Bedouin guides who would try to make sure that expeditions took as long as possible and required as much manpower as possible because they were being paid a daily rate. This kind of exploitation was a common experience for many travellers and most of them took it fairly gracefully. Freya Stark nevertheless put a lot of trust in her hired servants when it came to money and would let them carry her cash. She reasoned that to carry a strong box herself would attract too much attention. So the best solution was to hand over the responsibility to her personal servants who invariably responded well to the trust placed in them.

Even in the 1940s there was some personal danger for both Stark and Thesiger in travelling through Arabia. Thesiger wrote about passing towns where it was still regarded as a 'sacred duty' to kill Christians. Fortunately, the personal codes of honour held by their Bedouin guides kept most travellers safe. Once they had accepted the role of guides, the Bedouin believed that it was their responsibility to take care of their clients, even if that meant coming into conflict with other Bedouin. This might have been reassuring if it had not been for the fact that most Bedouin tribes seemed to be in a semi-permanent state of war with each other.

On several occasions, Thesiger was frustrated to find that his guides were not willing to go in a particular direction, because it meant entering the territory of another tribe with whom they were involved in a blood feud. The only way to overcome this kind of problem was to include a hostage from your enemy's tribe in your party, in order to keep their kinfolk at bay. This might sound strange, but it was a well-established system and the 'hostages' were often volunteers. Although they were much more war-like, you sense in the Bedouin a similar sort of pragmatism to that of the Inuit: meeting any outsider in the middle of the desert, as long as they weren't your direct enemy, was something positive. They might require your help but, equally, they might be able to help you. The Bedouin code of hospitality was aimed at creating a mutual support network: Thesiger might have sometimes found it galling to have to share his food, but on other occasions he was happy to enjoy the hospitality of strangers, when it was he and his guides who were hungry.

Unlike the Inuit or the Sherpas, the Bedouin didn't really benefit from their contact with explorers, because fundamentally they didn't really want to. Yes, they were pleased when Thesiger gave them new rifles, but no, they didn't want to adopt any of his values. Their society was too institutionally static and, to an extent, the Bedouin were proud of their isolation and made a virtue out of their harsh existence. Thesiger and Stark got a unique insight into Bedouin culture because they could speak Arabic but there was always an unbridgeable gap between the Bedouin and outsiders.

NATIVE NAMES FOR WESTERNERS
✱ ✱ ✱

Gringo Used all over South America to refer to white foreigners. There are several different theories about its etymology: it is sometimes claimed to come from the nineteenth-century song 'Green Grows the Grass oh' which was said to be constantly on the lips of visiting sailors. A more likely explanation is that it comes from the Spanish word 'griego', meaning Greek. This was synonymous with anything foreign and confusing.

Ferenghi Used all over Muslim Africa for white foreigners. It is thought to be a corruption of 'Frank', the ethnic name of the tribes which controlled Europe in the Dark Ages. The word was spread far and wide by Arab traders and variations of ferenghi are found in India, Thailand and even as far away as Samoa.

Gwailo Cantonese term for Westerner. It translates as 'white ghost' or 'ghost man'. Originally it was used to describe white sailors whose ships often left port at night or on the early morning tides, prompting the local people to think there was something supernatural about them.

Cheechaco Alaskan word for outsiders. In the nineteenth century, Alaskan settlers were scathing about anyone from the 'lower 48'.

Tuluk Nineteenth-century Eskimo word for Englishman, according to the French explorer Jean Malaurie; it is a corruption of the phrase 'to do'.

Kabloonak Term for white man used today in the Canadian Arctic; it means 'bushy eyebrows'.

Bwana A Swahili word used widely in Africa, it derives from an Arab word for father.

Sahib Hindi or Urdu respectful form of address, it derived from an Arabic word for 'master', 'lord'.

Umlungu Zulu word for white people, meaning 'people who practise magic'.

Tips for Travellers

IT'S GOOD TO TALK

The British explorer and adventurer Sebastian Snow spent many years in South America but he never mastered Spanish. Instead he used what he described as 'Good Loud English' to make himself understood. This approach got him through an ascent of Chimborazo and two trips down the Amazon but it is not really to be recommended. Learning to speak the local language obviously has huge advantages. One of the most effective explorers of Africa and the Middle East was Richard Burton. He was an expert linguist who was able to speak 25 languages and 15 additional dialects. Over a period of many years, he travelled widely throughout Africa and the Middle East and translated the *Kama Sutra* and several important Arabian erotic texts. Burton's linguistic skills were so effective that he was able to go in disguise into the holy cities of Mecca and Harra.

CHOOSE THE RIGHT GUIDES

In his classic book, *Scrambles amongst the Alps*, the British mountaineer, Edward Whymper, gave one of the most caustic assessments ever of the value of local mountain guides: 'They represented for me pointers of paths and large consumers of meat and drink but little more ... men whose faces expressed malice, pride, envy, hatred and roguery of every description, but who seemed to be destitute of all good qualities.' Some years later when he wrote a guide book for Zermatt and the Matterhorn, his advice to readers was somewhat more judicious: he warned them to keep away from old guides if they intended to do anything difficult, and told them not to hire anyone who had been involved in a lot of accidents or had a reputation for being a drinker.

The problem with Alpine guides in Whymper's day was that demand exceeded supply, so unsuitable men often put themselves forward. Even today you sometimes hear complaints about mountain guides who rush their clients or deliberately exhaust them early on, so that they don't have the energy to go all the way to the summit. If it is a guide's eightieth ascent of the Matterhorn or Mont Blanc then perhaps this is not so surprising. However, even though there are some rogues, the vast majority of guides are honest and in some situations indispensable to their clients. In fact, for all his occasional spite, Whymper was very generous to the guides that he respected and in a number of instances he was instrumental in setting up funds to help the families of guides who had died in the Alps.

RESPECT OF LOCAL RELIGIOUS VALUES

Whereas most of the early explorers of the Himalayas regarded the mountains as objects of beauty or physical challenge, for many of the local people they were sacred sites to be left untouched. A lot of British attention in the early days was directed towards Everest, which fortunately was not considered a terribly important mountain by Tibetan Buddhists. There are important monasteries on both sides of it, but Mount Kailas in western Tibet is considered a much more important mountain. In 1922 General Charles Bruce, leader of the second British Everest expedition, visited the head Lama of the Ronbuk monastery to inform him of their forthcoming attempt on the mountain. When asked why they wanted to climb it, Bruce replied that he and his team were pilgrims and that if they reached the top of Everest, they would be closer to heaven. Of course this wasn't quite true, but there was no malice in it. Thereafter, it became a custom for mountaineers to drop in at the monastery and request a blessing before heading up Everest.

In 1955 a British team made the first ascent of Kanchenjunga, the third-highest mountain in the world. It lies on the border of Nepal and Sikkim and is regarded as a very holy mountain by the Sikkimese. Before they made their attempt, the expedition leader, Charles Evans, visited the Maharajah of Sikkim and agreed that no one would actually tread on the summit itself. When some months later, two members of the British team, Joe Brown and George Band, reached the top, they stopped on a ledge which was 25 feet (7.5 metres) away from the summit peak, but barely a foot below it. They took a series of photographs with the small cone of snow behind them, and then descended.

Now you might say that visiting the Ronbuk monastery or protecting the summit of Kanchenjunga are just gestures, but respecting local traditions rarely costs very much and can add to the experience.

REMEMBER, FRATERNISATION IS
ALWAYS RISKY

Robert Peary described the presence of women, whether wives or native women, on expeditions as necessary to retain the 'top-notch of manhood'. If you go to the settlement of Quanaak in northern Greenland you can still find his descendants and those of his Afro–American assistant Matthew Henson. The temptation for explorers to fraternise with the locals has always been a strong one. Robert Peary believed that contact with women was essential to keep men healthy and vigorous, though he wasn't quite so forthright when his long-suffering wife Josephine turned up in the Arctic, only to find Peary's Eskimo lover Aleqasina pregnant and on board his ship. Having your spouse arrive unexpectedly is an obvious risk though not a very common one.

Misunderstanding local customs can be more of an issue; Peter Freuchen recorded an instance when he offered a lift on his sledge to an Eskimo girl. He didn't realise that this was tantamount to propositioning, nor did he understand why she refused and pointed to her hair. He later discovered that wearing her hair loose was a signal that she was menstruating, and therefore unavailable. Sexual etiquette is notoriously complicated, but love it seems knows no border. Freuchen later married an Eskimo and there were several African explorers who had native wives and mistresses.

HOW TO SAY 'WE COME IN PEACE'

In Arabia, Wilfred Thesiger noted that firing above someone's head was an inquisitive act rather than an aggressive one. If, in reply, an approaching stranger wanted to signal his friendly intentions, he picked up a handful of sand and let it fall between his fingers. However, when Captain Cook tried to get the attention of Australian Aborigines by ordering his men to shoot over their heads, they ran away into the bush, never to be seen again. Clearly there is no universal way of saying hello. Perhaps the best practice is simple persistence. When Meriwether Lewis met the Shoshone tribe on his epic trek across the American West, he had to try several gambits to prove that he came in peace. He spread a blanket on the grass and covered it with beads and trinkets; he called out 'tabba bone', the term for white man; when the first Shoshone ran away leaving their dogs behind he thought about tying a beaded handkerchief around their necks but the dogs wouldn't co-operate. Eventually Lewis tracked down an old woman and a young child who were unable to escape. He pulled up one of his shirt sleeves to reveal how white his arm was (in those days evidently, the Native Americans thought that white men were far less threatening than other natives) but he was by then so tanned from travelling, that he looked like a native from a different tribe. After giving them presents, he persuaded the women to take him to the rest of the tribe but before they set off, as a final gesture he and his two white companions anointed their faces with vermilion paint, yet another sign that they came in peace. Lewis may seem to have gone a little over the top, but his persistence paid off and soon he was greeting the chief.

DON'T FORGET TO TAKE A
MEDICAL HANDBOOK, OR A RADIO

When the Victorian traveller Mary Kingsley went to West Africa on an expedition to look for rare fish and fetish objects, she found herself pressed in service as an impromptu doctor. In one village alone, she had to deal with yaws, ulcers, infected limbs and worm-ridden eyeballs. She lanced the abscesses and attempted to clean the wounds but for most people there was little she could do. Fortunately, the villagers did not have overly high expectations of her. It is much more difficult when they do.

At the end of the of the *Kon-Tiki* expedition, Thor Heyerdahl and his crew met some Polynesians who took them back to their village and fêted them handsomely. Later, they were taken to a hut where they found a boy who had a large abscess on his head; they were asked if they could do anything for him. Heyerdahl's men still had some supplies of penicillin, but they were unsure of what dose to give and very concerned not to do more harm than good. Fortunately, they had a working radio transmitter and were able to get into contact with a doctor in the United States. He advised them to lance the abscess and then treat the boy with penicillin. It was a difficult operation made harder by some of the locals: the boy's mother became hysterical and some other villagers burst into the hut and had to be dragged away. It took two anxious days and more long-distance radio messages before the boy's temperature subsided, and several more before he was recovered. Heyerdahl breathed a deep sigh of relief; he had been quite worried about what would have happened if the boy had died.

HOW TO DEAL WITH HOSTILE NATIVES

The British explorer Henry Morton Stanley always travelled with a small army on his African explorations, so he was used to shooting his way out of trouble. Other explorers have had to be more resourceful. Sometimes doing the unexpected has proved to be the best thing.

The power of a photograph – When travelling through the Hadhramaut in Southern Arabia, Freya Stark was given a letter of introduction to a local dignitary. When she arrived at his house to call on him, he wasn't there and his family refused to welcome her, because of their vehement dislike of Christians. This kind of inhospitality was virtually unheard of and as Freya Stark walked away, she found herself being followed by a crowd of ill-wishers. They were growing more and more vociferous when Stark suddenly stopped, turned round, gathered them together and asked if she could take a photograph. They were nonplussed, especially as she seemed to be smiling at them. Her stratagem worked and they quietened down long enough for her to reach the edge of town and be reunited with her Bedouin guides. They hurried her along but Stark did not leave before she had taken a final photograph.

The power of a song – Whilst canoeing up a river on the border of Peru and Bolivia with a small party, the British explorer Percy Fawcett was attacked by local tribesmen. As the arrows rained down from the far side of the river, he and his men overturned their canoes on a sand bar and hid behind them. They were armed but Fawcett knew that he had too few men to fight his way out of trouble, and that even if he could it would mean the end of the expedition. So Fawcett told one of his assistants, a soldier called

Todd, to unpack his accordion and strike up a tune 'as though he was passing a jolly hour in an English pub'. Fawcett and the others joined in, singing whatever came into their heads. Gradually the arrows stopped coming and Fawcett noticed that one of their attackers had emerged from behind a bush. Fawcett then climbed gingerly into his canoe and crossed the river with two other men, still singing all the time. On the other side they were helped on to the bank by a large group of Guayaro braves. Soon they were invited to the local village where cordial relations were quickly established.

The power of passivity – Whilst the accounts of British explorers such as Samuel Baker, Richard Burton and Henry Stanley of their adventures in Africa are full of violent confrontations, the story of Mungo Park's first expedition to Africa is remarkable for its lack of blood and guts and for the apparent passivity of its protagonist. Mungo Park was a Scottish doctor who went to West Africa in 1795 at the behest of the grandly titled 'Association for the Promotion of Discovery through the Interior of Africa' on an expedition to trace the course of the Niger. He travelled with a small retinue of African servants, one of whom could speak English. From the beginning, Park was subject to numerous attacks and robberies by almost everyone he met along the trail: bandits stole his supplies of tobacco, the King of Bondon took his blue jacket, the nephew of the King of Kasson took much of his supplies; thereafter he was repeatedly shown off, examined, mocked and inspected. When he reached the kingdom of Ludamar he was held prisoner by Ali, its famous sovereign and forced to endure months of squalor and suffering. He eventually managed to escape and persevered with his journey until he finally reached the Niger in late July 1796. For all the privations of his journey, at his lowest ebb he always seemed to find a saviour. Once it was a slave who offered him a platter of nuts, at another time it was a woman who encountered him returning from her fields. She took him in for the night even though the local chief had banished him from his kingdom. Mungo Park managed to travel a short distance along the Niger but eventually, starving and penniless, his clothes in tatters, he was forced to retreat. His return journey was all the more difficult than the outward leg because he had to travel through the rainy season. Finally he returned to Britain after an epic journey which had taken him over two years and seven months. The amazing record of his adventures, *Travels into the Interior of Africa*, turned him into an instant celebrity though he told friends that he left out many of the more sensational moments for fear of stretching the public's credibility. Obviously, because he was travelling alone, it would have been very difficult for him to have offered any resistance to his various attackers so he had little choice but to be compliant. Nevertheless, the story of his journey is a testament to the power of passivity; as his journey went on and he was gradually being stripped of his possessions, it almost took on the quality of a Buddhist pilgrimage.

Wild Women

Although it is still the case that exploration and mountaineering are largely male domains, there have always been women who have been drawn to adventure and exploration. In the late nineteenth century there were many impressive women mountaineers; women such as Fanny Bullock Workman, Annie Peck and Elizabeth Hawkins Whitshed climbed to a very high standard in the mountains of Europe, South America and the Himalayas. Further afield, female explorers such as Mary Kingsley, Hester Stanhope and Gertrude Bell made exceptional journeys through Africa and the Middle East whilst in the early part of the twentieth century the aerial exploits of Amelia Earhart and Amy Johnson made them famous around the world.

Looking at it from a slightly different perspective, explorers' wives have often played significant roles in their husbands' careers. Women such as Kathleen Scott and Lady Jane Franklin were important standard bearers for their husbands, encouraging their ambitions and banging the drum for them in social circles. When their respective spouses died in the Arctic and Antarctic they both took an active role in moulding their legends. Lady Jane Franklin was a key figure in the opening up of the Arctic even though she didn't actually visit the region herself. After Sir John Franklin's expedition disappeared in the late 1840s, Lady Jane toured the world enlisting the support of presidents, prime ministers and emperors to finance several search expeditions. Neither Sir John nor his body ever returned but Lady Franklin became the first woman to be awarded a gold medal by the Royal Geographical Society in 1860 for her efforts. The various search expeditions brought new nations into Arctic exploration and opened up much previously unmapped territory.

It is not always easy being an explorer's wife and success and fame also have their downsides. After the 1953 Everest expedition, Joy Hunt used to long for the simple pleasure of being able to move through an airport without autograph hunters assailing her famous husband Sir John Hunt. Isabel Burton was besotted with her husband Sir Richard Burton, and eventually came to share his passion for all things exotic. She did not however share his interest in all things erotic and, a staunch Catholic, after his death she burned many of his papers, including a translation of *The Scented Garden*, a classic piece of oriental erotica. When news of her incineration leaked out, she received hate mail and was mocked in the press.

For any explorer's wife the main dilemma is whether to go with their husband or to stay behind. In many cases they have had no choice, but there are plenty of tales of honeymoons which mysteriously turned into trips to the Himalayas. Sitting at home waiting to hear whether your loved ones have survived their latest death-defying expedition and simultaneously running a household single-handed is not easy – but being dragged into the field by the bonds of love and devotion can be equally difficult.

Female Troubles

YOU CAN'T FIND THE RIGHT CLOTHES ...

For male explorers, clothing and equipment have often been an obsession; for women things have often been even more complicated. Whilst for men the issue is functionality, for women decorum is an added problem. Early female mountaineers frequently suffered from the tyranny of the skirt. Though eminently unsuitable for anyone climbing a mountain, in the nineteenth century most female climbers couldn't simply change into trousers. Mrs H.W. Cole, one of the first female alpinists, advised women interested in mountaineering to sew rings into the seams of their skirts through which a cord could be attached. Thus equipped they could raise their skirts instantly if the need arose. Mrs Aubrey Le Blond used to wear a skirt at the beginning of a climb and then abandon it for trousers as soon as she got higher up, away from prying eyes. Florence Baker, the redoubtable wife of the African explorer, Samuel Baker, was also a fan of trousers. As she and her husband journeyed through Egypt in the 1860s in search of the source of the Nile, she wore conventional clothing as they were passing through areas where they might encounter other Europeans and then switched to men's clothes when they were out of range. She was a skilled seamstress and ran up loose-fitting trousers and shirts for herself and her husband. Clothing for extremely low temperatures was a little less compromised; stockings were just as effective as trousers for keeping the cold at bay. Like their men folk though, winter clothing could be cumbersome: Josephine Peary's clothing list for her first expedition to Greenland in the 1890's was as follows:

1 knit kidney protector
1 Jaros combination suit
2 knit skirts
1 flannel wrapper
1 pair of knit stockings
1 pair of dearskin stockings
Snow shoes and a fur overall when she went outside

YOU CAN'T GET INTO THE CLUB ...

The formation of the Alpine Club in London in 1857 is regarded by many as the beginning of mountaineering as a sport. Initially membership required little more than optional attendance at an annual dinner and occasional citations in the club magazine, *The Alpine Journal*. It was nevertheless a prestigious club to belong to for anyone interested in climbing. Initially women were not allowed to join and there was little encouragement for them to take up mountaineering. It was fine to accompany their husbands and brothers on trips to the Alps but curtains twitched and eyebrows were raised if they attempted any climbing of their own. Of course many British women were not at all put off by this and by the end of the nineteenth century, climbers such as Lucy Walker, Katherine Richardson and the Pigeon sisters had climbed most of the high peaks of the Alps. In 1907, fifty years after the men got their club together, the Ladies' Alpine Club was founded in London with Mrs Aubrey Le Blond as its first president. Whether she wore trousers or a skirt at its first meeting is not recorded ...

YOU JOURNEY OUT TO MEET YOUR HUSBAND ONLY TO ENCOUNTER HIS PREGNANT LOVER ...

In the summer of 1901 Josephine Peary headed for Etah in the high Arctic hoping to meet her husband Robert, the famous polar explorer; he had been away from home for two years. Recently reports had come back that his work was going very badly and that he had been badly affected by frost-bite. When Josephine reached his base in Greenland, he wasn't there so the ship crossed the ice-clogged waters of Smith Sound to Payer Harbour on Ellesmere Island. He wasn't there either and before her ship could get away the mouth of the harbour became choked with ice, forcing Josephine and everyone on board to over-winter in the Arctic. Over the next couple of months she tried to persuade the local Eskimos to travel north to Robert Peary's other base at Fort Conger, hundreds of miles to the north, but no-one would take the risk. Amongst other things, she wanted to talk to him about an Eskimo girl, Aleqasina, that she had met at Payer Harbour, who was boasting that she was pregnant with the Great Peary's child.

Josephine Diebitsch was made of stern stuff. This was the third time that she had visited the Arctic. Three years after marrying Robert Peary in 1888, she had accompanied him on his first major expedition to Greenland. Though her friends warned her against going, she was terribly in love and found the hardest part of the expeditions were those when Peary was away in the interior ice cap. In her account of her year at Redcliffe House, the hut that they built in the far north of Greenland, Josephine revealed her conflicting emotions: she was fascinated by much that she saw but she was also terribly homesick and was squeamishly fearful of day-to-day life with the Eskimos.

It was clearly not an instant meeting of minds: when the Eskimos first encountered Josephine and Robert Peary, they asked each other which one was the woman. Josephine, for her part, was particularly frightened of encountering their fleas and described the Polar Eskimos as the 'queerest dirtiest looking individuals that I had ever seen'. In one very amusing section of her book, *My Arctic Journal*, she told the story of a 'Never-to-be-forgotten' night in an Eskimo igloo when she had had to put up with her hosts eating raw meat, stripping naked and picking fleas off each other. Nevertheless, for all her prudery, she had the good grace to recognise that the Eskimos often saw her as strange and amusing as she did them.

Meeting Aleqasina in 1901 stretched her devotion to the absolute limit but with tremendous forebearance she managed to forgive her husband and even spent time nursing Aleqasina through an illness. She had to wait nine months to see Robert Peary, but when she left the Arctic there was no sign of a rift between the two of them. Back in America, Josephine continued to use her social connections to lobby hard for him amongst 'the great and the good' and when setbacks undermined his confidence she was always there to support him.

Robert Peary finished his work in 1909, claiming to have reached the North Pole. His triumph, though, was clouded in controversy and though he was eventually given official recognition, for the next 11 years he was plagued by ill health. He died in 1920; Josephine kept on going until 1955. She had devoted most of her life to him and had subsumed her own ambitions, and her family life, to his obsession with the Arctic. In return she enjoyed a passionate marriage to someone who, for all his outward toughness, could be genuinely romantic and passionate. At one point early in his career, he had written that women were needed on expeditions in order for the men 'to remain contented'. Josephine was able to forgive Peary's infidelities, you suspect, because she was able to compartmentalise her husband's behaviour. At home he was devoted and faithful, even if he fathered two illegitimate sons in the Arctic.

YOU MARRIED THE WRONG EXPLORER . . .

It is hard not to feel sorry for Mary Moffat. When she met David Livingstone in 1843 he was new to Africa and she was the daughter of a famous missionary who Livingstone both looked up to and resented. In a letter to a colleague David Livingstone described Mary as 'a little thick, black-haired girl, sturdy and all I want'. Their's was not a marriage of star-crossed lovers; their courtship was brief and there was little romance.

Though Livingstone did not initially think of himself as primarily an explorer he was clearly ambitious to make a name for himself. In 1850 he, Mary and their three young children took a journey north towards the Zambesi River, even though she was pregnant. It was hard-going and they were frequently hungry and thirsty; Mary had her baby but it succumbed to a bronchial infection and died. This didn't put David Livingstone off though: in spite of opposition from Mary's parents, he took her on his next gruelling journey a couple of years later, even though she was again pregnant. This time they got through unscathed, only for Livingstone to then decide that the best place for Mary and the children was back in Britain. He promised her that she would be well looked after by the London Missionary Society but in the end she spent four miserable years being shunted around from relative to benefactor, living on meagre hand-outs from Livingstone's employers.

When the by now famous Livingstone returned to London several years later, he barely had time to see his family but, at her insistence, he agreed to take Mary and one of his sons back with him to Africa for an expedition to the Zambesi in 1858. This was a major, prestigious undertaking funded by the British Government. By now though, Mary Livingstone was a sad, broken woman who was fast becoming an alcoholic. When the ship reached the Cape, she realised that she was pregnant again and she left the ship to have her child at her parents' farm. Almost four years later she rejoined Livingstone who was still in the throes of his expedition. Just a few months later she came down with a severe attack of malaria and after a week of high fever, she died. In Livingstone's 600-page account of the expedition, her death was passed over in barely a page. She had perished, according to the good doctor, in the midst of a 'disinterested and dutiful' attempt to continue her work in Africa. He concluded 'Fiat Domine, voluntas tua' (Thy will be done). Livingstone's colleagues noticed that after Mary's death he softened somewhat and became more personable. It was all too late for Mary Moffat, who had died, feeling bitter and unloved, at the age of 41.

YOUR HUSBAND IS DEAD AND HIS BEST FRIEND SAYS THAT IT IS HIS OWN FAULT

Mina Hubbard was the wife of Leonidas Hubbard, a journalist working for a prestigious American outdoors magazine, *Outing*. In 1903 he and two companions, his best friend Dillon Wallace and a native guide George Elson, went to Labrador on an expedition to explore the Northwest and Naskapi rivers. Labrador was a vast wilderness on the east coast of Canada; it was home to several native tribes and there were trading posts along the coast but the interior was largely unmapped and unknown.

Leonidas Hubbard's expedition was a disaster: his party left too late in the season, took the wrong river north and tired themselves out carrying their canoes across difficult ground. They brought a small amount of supplies, having planned to live off local game but as they discovered, hunting was scarce and the weather was much worse than they had expected. Finally, after two months, they began to retreat, but they were by then so exhausted that one by one they began dropping off. George Elson managed to make it back to get help and Dillon Wallace was rescued but Leonidas Hubbard starved to death.

Back in New York, Wallace wrote a best-selling account of the expedition, initially with Mina Hubbard's blessing, but gradually they grew apart. There had been criticism in the press that the expedition had been badly prepared and it seemed to Mina that in his book Dillon Wallace had painted an unsympathetic picture of her husband and avoided his own responsibilities for what went wrong. So, when Wallace announced that he was going back to Labrador in 1905 to finish the job off, Mina Hubbard decided mount a rival expedition.

Mina had no experience of long expeditions but she had grown up on a farm in Ontario and knew how to fish and shoot. She had worked as a teacher and a nurse, and though she insisted on taking an air-mattress with her on her journey into the wilds of Labrador, she was hardy and tough. She had learnt how to use navigational instruments and was determined to both finish the journey and come back with a map. George Elson agreed to be her guide and with three other local men, they set off from a post on the North West River with two canoes and several weeks' worth of supplies in July 1905.

From the beginning, her journey went much more smoothly than her husband's, though 'smooth' is not quite the right word for a canoe journey along a river choked with rapids and boiling white water. In many respects Mina Hubbard was a passenger: the guides did all of the paddling, carrying, hunting and most of the cooking. George Elson in particular was very protective of Mina; having lost her husband on the previous expedition, he was determined that nothing should happen to her. Unsurprisingly, Mina found it 'somewhat irksome' to always be confined to camp whilst the others went out hunting and when she got an opportunity she took off by herself to explore. Her afternoon of freedom ended with a panicky chase around the woods after her men convinced themselves that she had been lost. After that they reached a compromise, allowing her to go off to take photographs whenever she needed to, but always in the company of a guide.

It was a painful experience retracing her husband's footsteps, compensated for only by the exhilaration of the journey and the sheer raw beauty of the landscape. She witnessed the epic migration of the Labrador caribou and was one of the first white women to encounter the Naskapi natives. In spite of her ever-protective guides, she still managed to feel like an explorer and was thrilled at the idea that she was discovering new things. Though the travelling wasn't as difficult as she had anticipated, nothing could have prepared her for the torment visited on everyone by the swarms of mosquitoes and flies. Inside her tent she was besieged by 'Labrador bulldogs' and 'snowstorms' of mosquitoes which beat on the walls and tried to find their way in. Outside, she sported home-made masks with visors made of netting but she could still feel the blood running down her neck when the mosquitoes attacked.

In the last few weeks, all her thoughts were on whether they would reach the settlement at Ungava in time for her to catch a ship back down south. She had given herself two months to travel 550 miles but in the closing stages it was getting very tight. If they missed the boat it would mean over-wintering in Labrador or retracing their route. The natives they met told them that they were far from the end but undeterred they pressed on. Finally, after five days shooting down the rapids at toboggan pace, they reached Ungava only to be told that the boat wasn't expected for another fortnight.

Though she didn't mention it in her account, *A Woman's Way Through Unknown Labrador*, she had not only succeeded in completing the journey that Leonidas Hubbard had proposed, but she had beaten Dillon Wallace to the end by six weeks. This was Mina Hubbard's only journey; her husband's memory had been honoured and she was able to draw up a map that would be used for years to come. After finishing her book, published under the name of 'Mrs Leonidas Hubbard', she went on a lecture tour. It eventually took her to England where she met her second husband. She died in 1953.

*   *   *

For both men and women, the human element is clearly a key factor in any expedition. Yes, equipment matters, but in the field a leader's most important resource is the other members on his or her team. Ultimately, luck plays a huge role in exploration and mountaineering, but good preparation and sensitive management make success much more likely. The next chapter looks in detail at six notable expeditions and analyses why some succeeded and some failed.

HOW TO ALWAYS HAVE A FRESHER PAIR OF UNDERPANTS ON A LONG EXPEDITION
* * *

1. Take two pairs of underpants or knickers, A + B
2. In week one, wear pair A
3. In weeks two and three, wear pair B
4. In weeks four and five, wear pair A which, having only been worn for one week, is now fresher than pair B
5. In weeks six and seven, wear pair B which, having only been worn for two weeks, is now the fresher pair
6. And so on, and so on . . .

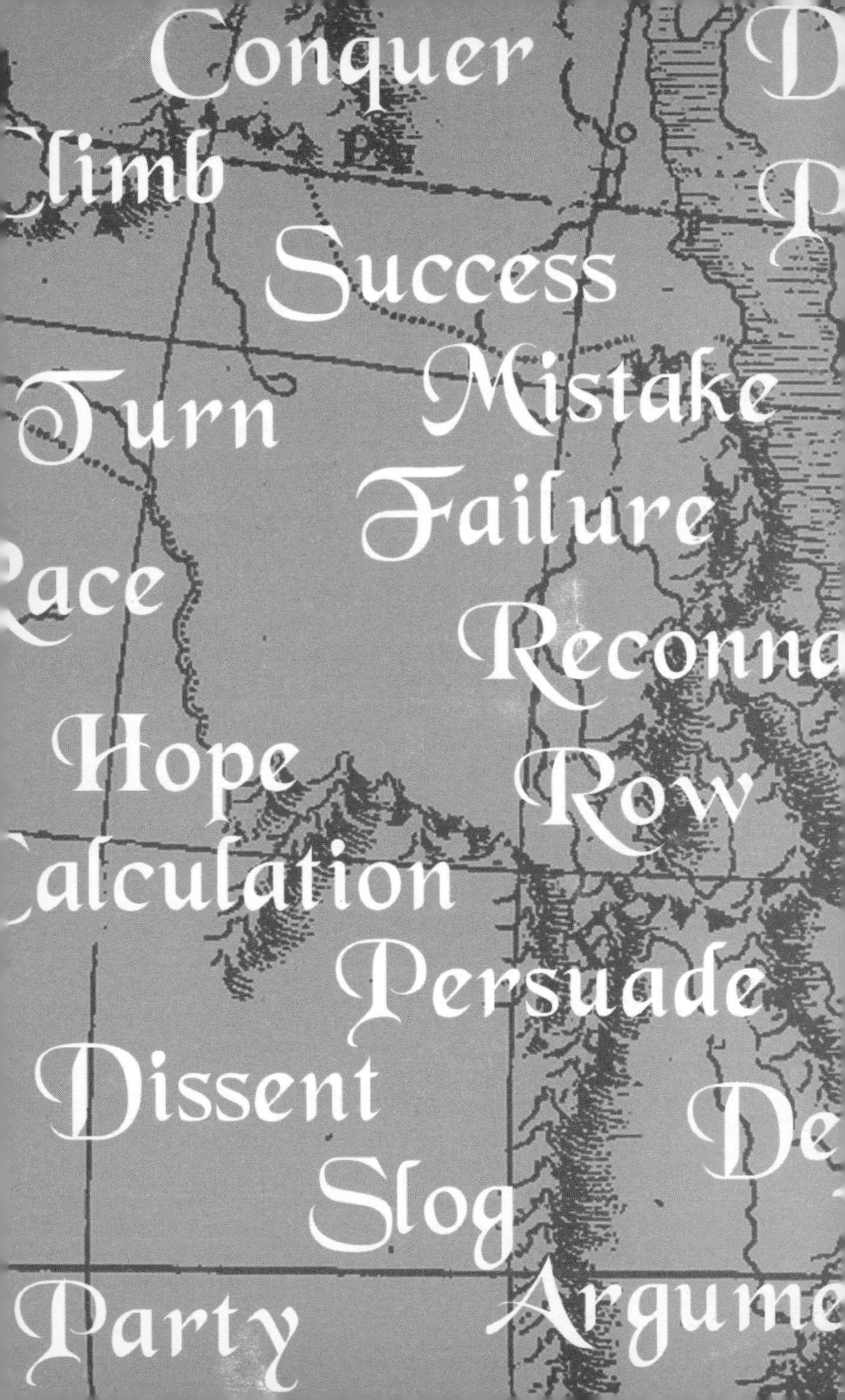

Conquer

Climb

D

Success

Turn

Mistake

Failure

Race

Reconna

Hope

Row

Calculation

Persuade

Dissent

De

Slog

Party

Argume

Getting There

I cannot say that the object of my life was obtained . . . I have never known a man to be placed in such a diametrically opposite position to the goal of his desires at that moment. The regions around the North Pole had attracted me since childhood and here I was at the South Pole. Can anything more topsy-turvy be imagined?

ROALD AMUNDSEN, ON REACHING
THE SOUTH POLE

Great God! This is an awful place and terrible enough for us to have laboured to it without the reward of priority.

ROBERT FALCON SCOTT, ON REACHING
THE SOUTH POLE AFTER AMUNDSEN

The one thing that John used to really hate was anyone saying that Everest had been conquered. A mountain is too big and too beautiful to ever be conquered. You should never call it a conquest.

JOY HUNT, WIFE OF SIR JOHN HUNT, LEADER OF
THE BRITISH EXPEDITION TO EVEREST IN 1953

What Makes an Expedition Succeed?

Exploration is a complex and inherently risky business. There are many more failed expeditions than there are successful ones. Sometimes failure is put down to luck and it is undoubtedly true that the natural world is rarely predictable; however, it is also the case that, to a large extent, explorers and adventurers make their own luck. Good preparation, a good team, a good leader and a good plan: none of these guarantees success but they make it much more likely.

When Hillary and Tenzing finally reached the summit of Everest in May 1953 it was the culmination of a superbly planned expedition. It was well funded and carefully organised. Unlike so many previous Himalayan adventures, there were no fatalities and very few injuries. Fifty years on Everest has lost some of its lustre, but this expedition remains a benchmark for Himalayan mountaineering. Above all, its success was down to the leadership of John Hunt and the teamwork of all the expedition members.

BACKGROUND

Before the war, Britain had a monopoly on Everest because of its political influence in the region. Seven British expeditions went out to Tibet to try to climb the world's highest mountain, and though some had come close,

EVEREST: THE SAGA BEGINS
✷ ✷ ✷

1849–50	Peak XV is measured by the British Survey of India
1856	Peak XV is computed to be the highest mountain in the Himalaya and named 'Everest' in honour of Sir George Everest, a former Director of the Survey of India. Its height is estimated to be 29,002 feet (8,700 metres)
1921	British Expedition travels to Everest via Tibet to make a detailed reconnaissance of the mountain
1924	Mallory and Irvine disappear high on Everest
	The Tibetan government and the Dalai Lama take offence at the 1924 expedition film. There are no further expeditions for another six years
1933	Frank Smythe, Percy Wyn Harris and Laurence Wager reach 28,000 feet (8,400 metres)
	The Houston expedition makes the first flight over Everest
1934	English eccentric Maurice Wilson flies halfway round the world in his airplane *Ever Wrest* to make an abortive solo attempt. His emaciated body is found a year later.
1945–47	RAF makes secret flights over Everest
1950	Tibet closes its borders to foreigners; Nepal opens hers
1951	Eric Shipton leads a British reconnaissance expedition to Everest via Nepal. To their surprise the British team find out that the Nepalese government have given a Swiss team the permit to climb in 1952
1952	Two Swiss expeditions make attempts on Everest from Nepal. They are the first non-British climbers on the mountain. Raymond Lambert and Sherpa Tenzing reach 28,220 feet (8, 466 metres)
	British climbers get permission to make an attempt from Nepal in 1953

none succeeded. After the war everything changed. Tibet was invaded by China and the new authorities barred any further expeditions. At almost the same time, the previously closed kingdom of Nepal opened its borders, thus enabling climbers to approach Everest from the south. Britain lost its monopoly on Everest. In 1952 a Swiss team almost made it to the summit, and when a British team obtained a mountaineering permit from the Nepalese government for the next attempt, they knew that if they failed in 1953 Everest was 'booked up' for the next few years and that another country would most likely win the greatest prize in mountaineering. It was now or never, so nothing was left to chance.

THE LEADER

The 1953 expedition got off to an inauspicious start when the organising committee manoeuvred to get rid of Eric Shipton, the hugely respected mountaineer who was due to lead the team. His replacement was Col John Hunt, a serving officer and mountain warfare specialist. It was a very awkward way for him to start, but Hunt turned out to be a very effective leader. Before they met him, a lot of the other climbers were worried about his military background, but once they had, they were all won over by his disarming charm and enthusiasm. Hunt was both charismatic and well practised in the art of leadership; some of the team joked that he behaved like someone who had read *How to Win Friends and Influence People*. He understood how important it was to take a personal interest in all the men and to make them feel special. His other great strength was his careful approach to logistics and planning; before leaving for Everest, Hunt drew up detailed plans for almost every aspect of the expedition but on the mountain he was flexible enough to adapt to prevailing circumstances.

THE TEAM

Hunt's first task was to bring the men already chosen for the expedition back on side. Several of them had gone with Eric Shipton in the previous year to Cho Oyu and they weren't happy about his dismissal. One man, Tom Bourdillon, formally resigned in protest. Over the next few months, Hunt went to see the men individually and, slowly but surely, he won them round. He took on two new members, George Band and Mike Westmacott; both were young and promising. Hunt didn't meet the two New Zealanders who had also been on Shipton's expedition , George Lowe and Edmund Hillary, until he reached India; initially he wasn't sure that he wanted to take anyone whom he hadn't met, but the other British climbers insisted that they were vital. Hunt followed the previous year's Swiss team, inviting the Sherpa Tenzing Norgay to become part of the climbing team. There were tensions though at the beginning of the expedition: Tenzing wasn't sure that he wanted to go with the British. He had struck up a great friendship with the Swiss climber Raymond Lambert in 1952, and hoped to make another attempt with him. The local press gleefully reported that some of the Sherpas weren't happy with the accommodation they were put in at the British embassy in Kathmandu; later one Sherpa stormed off, complaining that the British were being mean about distributing expedition clothing. These, though, were minor issues; as the expedition trekked in to Everest, the Sherpas and men began to bond into a strong, cohesive team. Hunt later wrote that one of his fondest memories of the expedition was the moment when Tom Bourdillon turned to him and said what a happy party they were.

EQUIPMENT AND PREPARATION

Another reason for the expedition's success was the time and effort that Hunt and the other men put into finding the best clothing and equipment. They spent months scouring Britain and Europe, experimenting with what they had found. For items such as ice axes and crampons they opted for the tried and tested, but for other pieces of equipment they went right back to the drawing board. For example, they took no less than nine different materials to an RAF wind tunnel to choose the best fabric for their tents and outer clothing. Even more obsessive was their search for the perfect high-altitude boot; after months of testing, the final design was manufactured using the combined might of no less than 33 British shoe companies. On Everest they discovered that some items were not perfect but there's no doubt that this expedition set new standards for equipment and preparation.

OXYGEN

Hunt's team approached the oxygen question thoughtfully. Pre-war expeditions had often been ambiguous in their attitude to oxygen, but the 1953 team had no doubt that Everest would not be climbed without it. When the expedition was announced, John Hunt was inundated with letters from madcap inventors who suggested everything from laying an enormous oxygen pipeline up the Lhotse Face to firing oxygen cylinders up to the South Col with a huge mortar. More practically, Hunt's expedition benefited from all the work that had gone on during the Second World War into developing lightweight gas canisters, a vital issue for equipment which would have to be carried so high on the mountain. Tom Bourdillon was the team's oxygen expert; he designed and built a special type of set which was much more efficient than the standard apparatus. In 1953 they took two of Bourdillon's sets along in addition to several standard models.

The expedition also gained significantly from the research done by a British physiologist, Dr Griffith Pugh. He had carried out a lot of field-work on the British expedition to Cho Oyu in 1952 and continued the research in his laboratory in London. Pugh spent many months working out the optimum flow rates for oxygen sets, the volume of liquid that climbers needed to drink at high altitude and how much food they needed for optimum performance. No other expedition had ever had the benefit of this level of research, and ultimately it did make an enormous difference.

THE EXPEDITION

In March 1953 Hunt's party arrived in Nepal. At the time, Britain was one of the few countries with an embassy there and the ambassador was able to offer them a base and a lot of logistical support. A month later, they reached the foot of Everest and set up their base camp. Their first task was to get through the Khumbu icefall, a tangled mass of crevasses and seracs at the entrance to the Western Cwm. It is one of the most dangerous parts of the route up Everest and the previous expeditions had had a lot of problems getting through. In order to bridge the crevasses they brought aluminium builders' ladders, initiating a practice that has continued up to this day. Then Hunt split the men into small teams to tackle different stages of the route up the mountain. After a month he brought everyone together to announce his plan for the second stage of the climb. He chose two pairs to make the final attempt on the summit, but he made sure that everyone else realised that they too had a vital role to play in the success of the expedition. Some men were disappointed not to be chosen for the final push but they all swallowed their pride and got on with it.

Hunt recognised that one of the main reasons for the failure of the two Swiss expeditions of 1952 was that they had been unable to get enough supplies up the Lhotse Face, a huge slope which had to be negotiated before any attempt on the summit. When Tenzing and Raymond Lambert made their first attempt in May 1952, they didn't even have a sleeping bag to share between them because the Swiss team had such problems on the Lhotse face. A key part of Hunt's plan was to establish a well-stocked camp at the top of it on the South Col. This wasn't easy though and Hunt had to throw much more effort at it than he had anticipated. George Lowe led this section of the climb: he spent ten painstaking days hacking a staircase of steps up the steep slopes of the Lhotse Face. Until a safe route was prepared, the Sherpa porters wouldn't be able to ascend with their loads to establish the next camp. Several other climbers were sent up to help Lowe, but most of them were so affected by the altitude that they did not last long. Finally Hunt agreed to allow Hillary and Tenzing to take part in the effort; previously he had hoped to keep them back in reserve but Tenzing persuaded him that his presence would inspire the other Sherpas. Sure enough they reached the top of the Lhotse Face carrying hundreds of pounds of supplies and equipment and set up camp on the South Col, a windblown ridge which separated the Lhotse Face to the south from the Kanshung Face to the east. Now they were ready for the final stage of the campaign.

The delays on the Lhotse Face were a major worry and the weather was often poor but, by and large, it was going well. Hunt had thought it all out in advance and on the mountain he was confident enough to alter his plans when he needed to. Two of his great strengths were preparation and pragmatism and his choice of summit teams showed this. He chose Charles Evans and Tom Bourdillon to make the first attempt; it was principally intended as a reconnaissance but he told the men to carry on to the top, if on the day the felt that they were capable of it. In the event, problems with Charles Evans' oxygen set persuaded the two men to retreat but not before they had broken Tenzing and Lambert's altitude record and reached the South Summit, a stump of rock 300 feet (100 metres) below the summit.

Bourdillon and Evans could have tried to go further but, if anything had gone wrong, they knew that it would have put the whole expedition in jeopardy. They didn't want Hillary and Tenzing's attempt to turn into a rescue mission so they selflessly turned around. Bourdillon regretted it later, but at the time it was the most team-minded thing to do and having reached the South Summit, they had useful advice to pass on to the next pair.

It was now Hillary and Tenzing's turn. Right from the beginning they had shown themselves to be two of the strongest climbers in the team. Both men were team players but Hunt could also sense that they were both hugely competitive and very ambitious. The fact that one of them was a New Zealander and the other a Sherpa didn't matter; getting at least two members of the team to the top was more important than flying the flag. Hunt liked the idea that one of the summit pair should be a Sherpa because of the huge contribution they had made to Himalayan climbing, but this wasn't a sentimental decision: Tenzing was not as technically accomplished as Hillary but he had huge stamina and strength and had proved his mettle on the Swiss expeditions.

In order to give Hillary and Tenzing the best possible start, it was decided to set up another camp further up the summit ridge from the South Col. Hunt and one of the Sherpas went up with a tent and some spare oxygen cylinders on the same day that Bourdillon and Evans made their attempt. When Hunt and Da Nymgal returned, they were both absolutely exhausted. Hunt had decided earlier on that because of his role as the leader, he would not take part in the summit attempts, but he wanted to stay on the South Col to be close to the heart of the action. However, when Tom Bourdillon and Charles Evans came down, a problem arose. They were too weak to carry on to base camp by themselves so Hunt asked George Lowe to accompany them. Lowe didn't want to go; he pointed out in no uncertain terms that Hunt should go down. Lowe's argument was that at this stage, he was fitter and stronger than Hunt and would be of more use if anything went wrong. Hunt refused to budge and George Lowe reluctantly prepared to descend. Then, after some thought, Hunt changed his mind and accepted that Lowe was right; he should be the one to descend. 'I have never admired John more than I did at that moment,' Hillary later wrote in his memoirs. John Hunt described it as one of the worst points on the expedition. It may all seem like a small matter, but this episode summed up Hunt's whole approach to the expedition. Even though he was the leader, the one who should be giving the orders, he was willing to be persuaded by sound arguments to put aside his own ambitions and desires for the good of the team.

On 29 May, Hillary and Tenzing set off for the summit. Following Griffith Pugh's advice, they slept with their oxygen sets switched on and drank copious quantities of lemon tea before breaking camp. On the way up, Hillary was constantly checking their oxygen cylinders to make sure they weren't using too much. The most serious obstacle that they met was a 40-foot (12-metre) rock face fringed with ice; down below, it would not have been a problem but at 29,000 feet (8,700 metres), it was quite a challenge. In honour of its first ascent, it has since been known as the Hillary Step. When they reached the summit, Hillary had enough time to take several photographs of Tenzing and to search for any signs that Mallory and Irvine might have preceded them. He found nothing. The two men descended carefully to the South Col and then on the next day they went down to advance climbed base camp to celebrate with Hunt and the other members of the team. Three days later, on the night before the coronation of Queen Elizabeth II, the news reached London.

Technically, the Swiss team which failed in the previous year were better climbers than the British; several of them were professional guides and most of them had kept on climbing during the Second World War. So why did Hunt's team succeed? Obviously there were lots of factors involved, but it crucially comes down to organisation and teamwork. There were so many different climbers who played an important role in getting Hillary and Tenzing to a position where they could make a serious attempt: Wilfred Noyce who was the first to break through on to the South Col, Charles Wylie who looked after the Sherpas so carefully, and George Lowe who did most of the work on the Lhotse Face, to name but a few. At the head of all of them there was John Hunt, carefully getting the best out of each mountaineer and Sherpa. He was a modest man and, for all his military background, he hated anyone referring to his 'conquest' of Everest. This was just one attempt and as he well knew, no mountain could ever be conquered.

A SHORT HISTORY OF POLYNESIA
✦ ✦ ✦

1520	Magellan 'discovers' the Pacific
1595	Spanish sailors land on the Marquesa islands
1767	Captain Samuel Wallis lands on Tahiti on HMS *Dolphin*. He raises the Union Jack and claims the island for Britain. His sailors trade nails from the ship for sex with the local women
1768	French explorer Bougainville lands on Tahiti and claims the island for France
1769	Captain James Cook arrives on Tahiti aboard *Endeavour*. He builds an observatory at Mahina Point to record the transit of Venus
1789	After setting Captain Bligh adrift, mutineers from the *Bounty* land on Tahiti. Sixteen decide to stay
1797	Arrival of the first missionaries from the *London Missionary Society*. They remain until 1852
1790s–1860s	Native population declines by 80–90% after epidemics of TB, typhoid, influenza and smallpox
1843	Tahiti becomes a French protectorate
1880	Tahiti officially becomes a French colony
1891	Gaugin, the impressionist painter, arrives on Tahiti. He dies there in 1903
1937	Thor Heyerdahl, a young Norwegian zoologist, spends his honeymoon on the island of Fatu-Hiva with his wife Liv. They want to see whether they can go back to nature, but after two years feel an 'inconvenient urge' to return to civilisation

The Kon-Tiki Expedition

The 1953 British Everest expedition was the culmination of a long history of effort and energy directed towards the solution of a particular problem, namely getting up the highest mountain in the world. Thor Heyerdahl's *Kon-Tiki* expedition by contrast came entirely out of the blue. He and his small team of Scandinavian adventurers had no history of similar endeavours to study and improve upon; their inspiration came from ancient accounts of primitive sea craft and the legends of Polynesian islands.

Unlike much exploration, the purpose of the expedition was not to discover new territory or claim some geographical prize. Heyerdahl's aim was to prove a theory of his: that some of the early settlers of Polynesia could have come from South America. The academics to whom he had proposed this were either sceptical or dismissive. It wasn't possible, they said, because the ancient South Americans didn't know how to build ships and it was plainly impossible to travel 4,300 miles (6920 kilometres) on the balsa wood rafts that they were known to have used. Heyerdahl demurred; he had read a lot of early accounts of the Spanish conquistadors and seen references to large rafts being sailed along the coast, and he was convinced that his theory was correct. So he decided to prove his detractors wrong by building a raft and sailing it across the Pacific.

This expedition is notable for its sheer daring. If the raft had capsized in the middle of the Pacific Ocean there would have been virtually no chance of anyone surviving. It was a bold gesture from start to finish, which bordered on the reckless, but managed to stay on the right side of the line because of the serious way in which it was carried out.

THE LEADER

Heyerdahl had trained as a zoologist. His first taste of adventure came in 1937 when he went to the remote Polynesian island of Fatu-Hiva, aiming to find out whether he and his wife could 'go back to nature'. It was much harder than they had expected but they managed to stay there for over a year. On Fatu-Hiva he heard stories from village elders of how their ancient god, Tiki, originally came across the sea from the place where the sun rose. Could this mean that they came from South America? It lay to the east of Fatu-Hiva and Heyerdahl had observed that the strongest ocean currents flowed east to west. He had also noticed that some of the statues found in Polynesia and on Easter Island resembled artefacts discovered in South America.

Heyerdahl published the first article on his theory about South American migration in 1941 whilst working at the Museum of British Columbia. His academic career was interrupted by a return to Europe for a stint in a parachute regiment of the Free Norwegian forces, but when the Second World War ended Heyerdahl headed for New York where he tried to convince Polynesian specialists to take his theory seriously. They didn't, so he came up with the *Kon-Tiki* expedition.

Heyerdahl was one of the few explorers who really could be called visionary. He was somewhat of a showman and later he sometimes seemed deliberately to court controversy, but there always seemed to be an idea behind the bravado. The living history approach of the *Kon-Tiki* expedition is commonplace today, but in 1947 the idea of sailing a balsa-wood raft across the Pacific was considered utterly foolish. Heyerdahl's leadership skill was as much in the conception of the project as it was in its execution.

THE TEAM

Heyerdahl decided that six was the ideal number for the crew: it divided the day neatly into 4-hour watches and was a sufficient number to offer what he called a 'a change of society'. Five of the team were Norwegian, the sixth a Swedish ethnologist, Bengt Danielsson, one of the few scientists to take Heyerdahl's theory seriously. Erik Hesselberg was an experienced sailor but the rest were landlubbers; Heyerdahl didn't want to give himself an unfair advantage over the ancient Peruvians by filling his raft with naval men. Knut Haugland and Torstein Raby had been wireless operators behind enemy lines during the Second World War; Herman Watzinger was an engineer who gave up his job to join the expedition. They were all tough, self-reliant and used to being on their own, the kind of men who wouldn't lose their cool in a crisis. And they needed to be, for this was a dangerous, 'all or nothing' experiment. The whole enterprise was invested with a kind of wartime spirit; Heyerdahl, Raby and Haugland knew what it was like to put their lives at risk. In the heyday of polar exploration, Scandinavian explorers such as Nansen, Amundsen and Nordenskjold showed the time-honoured Viking spirit in their daring adventures. Heyerdahl's team had a raft rather than a longboat, but the same spirit prevailed.

EQUIPMENT

Though it wasn't a hugely expensive expedition, Heyerdahl had a lot of trouble raising funds. He had to make up for his lack of cash through his networking skills. At the Explorer's Club, Heyerdahl was befriended by Peter Freuchen; Freuchen helped raise the profile of the expedition and there were soon offers of funding from the press. Ultimately these didn't materialise but Heyerdahl fared better with the military; he was able to persuade US Air Command and the British and American armies to part with a lot of equipment, in return for agreeing to test its seaworthiness. His booty included sunburn cream, splash-proof sleeping bags, floating knives, shark repellent and 684 tins of pineapple chunks.

The construction of the raft was a saga in itself. Heyerdahl went to Ecuador to buy the wood but when he arrived, he was told that there were no suitable balsa trees available and that he would have to wait for six months. This would have ruined his plans so Heyerdahl took drastic measures. He and another member of his team armed themselves to the teeth and, accompanied by an Ecuadorian officer, headed inland to a remote settlement on the far side of the Andes. They found a remote plantation and persuaded the owner to have 12 trees cut down for them. These were duly rolled into the water and then floated downriver, accompanied by the two Norwegians and some local men on a smaller raft. Then Heyerdahl persuaded the president of Peru to allow the *Kon-Tiki* to be built in the main naval dockyard in Callao. It was all *Boys' Own* stuff; at one stage Heyerdahl had even offered to parachute into Ecuador to get the balsa wood if there was no other way. The simple fact was that Heyerdahl was good at making things happen. He had no expedition committee behind him to shoulder the burden; it was all down to one leader and a small team.

The *Kon-Tiki*'s design was based on records of ancient Peruvian craft: it had a 19-foot (5.7-metre) steering oar, a small cabin made of bamboo and a mangrove-wood mast which carried a large canvas sail with a painting of Tiki on it. On board there were four months' worth of military rations and 250 gallons (1137 litres) of drinking water. Each man was allowed a minimum of personal possessions; Bengt Danielsson, the Swedish ethnologist, elected to fill his quota with 73 sociology and ethnology books. None of the many visitors who came to see them had any faith in the raft; some placed bets on how long it would last in the water, while another donated a Bible to help them prepare for the inevitable. This first stage of the expedition showed Heyerdahl at his best: daring, charming and utterly focused.

SETTING SAIL

On 28 April 1947, the six men left Callao harbour escorted by a gaggle of photographers and a small ship from the Peruvian Navy. The *Kon-Tiki* survived its early encounters with big seas, though the men found that they had to rig up elaborate systems of ropes in order to be able to handle their long and extremely heavy steering oar. The obvious advantage of a raft over a boat was that all the tonnes of water that came on board simply flowed out through the gaps in the logs and never came back in again. The big disadvantage was that the *Kon-Tiki* was very difficult to steer and if anyone fell overboard it would be virtually impossible to turn round.

At first there was plenty to worry about. The raft was at the mercy of the currents and the winds, which for the first few weeks threatened to send it north to the Galapagos Islands. And that wasn't all: with alarm, they watched the balsa-wood logs beginning to absorb water; would the raft sink as their detractors had claimed? They were also worried that the hemp ropes, which lashed the raft together, would simply wear out because the logs seemed to be in constant motion. Luckily the raft kept a westward course and only the outer few inches of the thick balsa logs became saturated with water. The ropes gradually became embedded in this spongy outer layer, protecting them against the salt water and the constant chafing.

They tried to send back regular meteorological bulletins but for the two wireless operators it was a battle just to keep their apparatus working. One of their biggest challenges was to find a way to get their aerial high enough; they experimented with tying it first to a balloon and then to a kite, but neither trick worked so they had to carry on experimenting. The sociologist read his books, the artist sketched, one man looked after the raft and Heyerdahl kept the ship's log and filmed life on board. The 'dirty' jobs of cooking and steering were evenly split and if there was any important decision to be taken, they usually had a group discussion.

Life on board was a mixture of languid monotony interspersed with occasional bursts of frenetic activity. The numerous fish that swam around the *Kon-Tiki* helped keep boredom at bay and provided them with fresh food. During the day they sometimes encountered schools of whales and once a huge 50-foot (15-metre) whale shark circled the raft, full of menace and playfulness. One man thrust a harpoon into it, but the monster from the depths snapped it like a twig. Later they worked out how to capture small sharks, which became a welcome addition to their diet. Two men restricted themselves to the army rations, while the others had a more varied diet with lots of fish and plankton, which they caught in a silk net attached to the back of the raft. After two months, their freshwater supplies turned stale but they were able to collect ample rainwater.

On day 45 they found themselves halfway through their journey with another 2,000 miles (3,219 kilometres) to go. They noticed a mysterious mid-ocean reef on their chart and altered course to investigate it, but they found nothing. A minor panic ensued when they tried out a rubber dinghy given to them by US Air Command; once it was in the water, it couldn't keep pace with the *Kon-Tiki*, even with the sail down. It took them several panicky hours trying to get the dinghy and its crew back on board.

None of them had known each other for more than a few months before they joined the expedition but their temperaments were evenly matched and, as Heyerdahl calculated, with six men on board it took a while before they got bored of each other. One way of looking at the expedition is to see it almost like a social experiment or, dare I say it, a prototypical piece of reality TV. Episode one, find a team; episode two, build a strange craft; episodes three to six, try to sail it across the Pacific Ocean and see if everyone gets out alive.

It is interesting how small details of the raft's construction had a

psychological impact. In his journals, Heyerdahl noted how the bamboo cabin in the middle of the raft gave them a real feeling of security precisely because it looked and smelled so out of place in the middle of the sea. Though the raft was barely 19 x 45 feet (6 x 14 metres), Heyerdahl had been conscious of trying to create some sense of separation between the various spaces in order to make life on board just a little bit less monotonous. So some of the raft had wicker decking, some bamboo and some nothing at all.

On 30 July they saw land for the first time in three months, a small island called Puka Puka, the first of the Tuomotu group. Heyerdahl had been vindicated, but he wasn't ready to celebrate yet. It would be terribly bad luck but, because it was so difficult to steer, there was a real danger that the *Kon-Tiki* might sail through Polynesia without touching land at all! And indeed, the *Kon-Tiki* sailed right on past Puka Puka. Four days later the men put their flags up as they approached the island of Angatu; according to their charts it was surrounded by reefs which would make it difficult to land. Suddenly they passed a village near the shore. A canoe was launched and before long there were two Polynesians on board cadging cigarettes and repeatedly saying 'Goodnight, goodnight' because it was the only English they knew. When their guests realised that the *Kon-Tiki* didn't have an engine and that the men were going to try to paddle ashore, they scarpered.

Heyerdahl thought that they had been abandoned until four more canoes arrived. They tied ropes to the raft and attempted to tow it ashore. By now it was early evening and a strong wind was blowing from the island out to sea, and though everyone started enthusiastically the combined might of the four canoes and the *Kon-Tiki*'s own rubber raft wasn't enough. Then to compound their problems, Knut Haugland misunderstood a message from Heyerdahl and went ashore with one of the canoes. Meanwhile the *Kon-Tiki* continued to float away from the island. It was all going wrong.

Heyerdahl had almost given up hope of seeing him again, when suddenly Haugland appeared in another canoe and scampered aboard. Somehow he had persuaded the islanders to take him back out, regardless of the risk. The whole episode had a slightly comical quality about it, but it also showed the men's ability to keep their cool in a crisis. In spite of the fact that neither Haugland nor the Polynesians could speak each other's language, he had managed to talk them into setting off for a third time from their village into the dark of the night.

The canoe left and the raft sailed restlessly on. Directly in their path lay the Raoria reef; behind it there was a series of small palm-fringed islands. It was a known hazard for shipping and, as Heyerdahl realised, it was almost inevitable that they would crash into it. They put anything valuable into watertight bags, and readied themselves to be shipwrecked. Every man knew his job and they all got on with it as calmly and quietly as possible. As they got closer, they could see the wreck of a previous ship lying in the midst of the reef. 'Order number one,' as Heyerdahl put it, was to hold on to the raft whatever happened; at some point, it would inevitably get to the shore. One of the wireless operators, Torstein, hammered out a final message to a radio enthusiast with whom he had recently made contact.

Then the raft was over the reef, the waves were boiling, the breakers were crashing and the men found themselves being thrown up and down as the *Kon-Tiki* was repeatedly sucked into the water and then forced to the surface. At various points they were all thrown overboard, but the ropes held fast and everyone came out the other side. The raft was caught on the reef but the men swam and waded ashore where they joyfully sank their toes into the welcoming sand. A few days later two Polynesians arrived from a nearby village. After numerous celebrations, Heyerdahl and his men headed first for Tahiti and then back to Norway.

The voyage of the *Kon-Tiki* was reported all around the world and soon Heyerdahl was being celebrated as one of the greatest explorers of his time. The success of the expedition, however, didn't win all the experts over to Heyerdahl's migration theory. For many years afterwards he maintained a running battle with what he saw as the small-minded world of academia. Today most experts agree that Polynesia was in fact peopled by Asians from the west and not South Americans from the east and cite new linguistic evidence to prove it. In his defence, Heyerdahl never claimed that the *Kon-Tiki* expedition would prove that ancient South Americans *had* crossed the water to Polynesia; all he wanted to show was that they *could* have done so. The *Kon-Tiki* had survived the journey, the winds and the ocean currents had taken them in the right direction and the voyage had been carried out by an inexperienced crew. Anyone who claimed that it was simply impossible to take a balsa-wood raft across the Pacific was proved wrong.

The voyage of the *Kon-Tiki* was a pioneering expedition which inspired many others to follow in its wake. It was a hazardous undertaking carried out with modesty and understatement by a very cool-headed crew. The whole concept was very dangerous but the risks were balanced by careful and thoughtful research. Heyerdahl picked the right companions, chose his equipment intelligently and refused to be thrown off course by any problems that came his way. Years later he made further voyages in the *Ra*, a papyrus boat built to find out whether Ancient Egyptians could have crossed the Atlantic, and the *Tigris,* a reed boat which was built to test the navigational skills of the ancient Sumerians, but neither of these had the same international impact.

The Search for the Source of the Nile

Length : 4,184 miles (6,733 kilometres)
Route: The Nile flows through nine countries
Cities close to the Nile: Cairo, Gondokoro, Khartoum, Aswan,
Luxor, Karnak, Alexandria
Tributaries: The White Nile, the Blue Nile, the Atbara

In 1864 one of the great controversies in Victorian exploration reached a climax at a meeting of the British Association in Bath. On Friday 16 September, the crowds turned out for a debate between Richard Burton and John Hanning Speke. The two men were former colleagues, now turned bitter enemies. Speke had claimed to have discovered the source of the Nile in East Africa, but Burton had vociferously and publicly doubted him and poured scorn on his so-called 'proofs'. Speke had replied in kind, denigrating Burton's abilities as an explorer and questioning his personal morality. Now the two men were getting together for a very public showdown. It was rumoured that Speke was literally planning to kick Burton off the podium. However, the public were denied their fisticuffs. On that morning, after a long delay that set the crowd on edge, the president of the Royal Geographical Society announced that Speke was dead. He had been killed on the previous afternoon in a mysterious accident whilst out hunting. Burton quipped in a letter to a friend 'the charitable say that he shot himself, the uncharitable that I shot him', but the truth was that he too was devastated by the events.

The seeds of this brutal finale had been sown eight years earlier when the two men set off for East Africa in search of a huge lake which some believed might be the source of the Nile. The expedition that followed was both wildly successful and hugely flawed. On the one hand it was a story of incredible bravery and physical stamina, while on the other it offered a powerful warning on the dangers of ambition and egotism.

THE LEADER

By the late 1850s, Richard Burton was one of Britain's best-known Orientalists and explorers. His fame principally rested on his trip to Mecca disguised as an Afghani doctor. He was only the second European to have entered the holy city and the 2-volume account of that journey impressed both the scholars and the public. He followed it with an equally dangerous expedition to the holy city of Harrar in Ethiopia; this time he entered the city dressed as a British officer but the risks were just as great. For his expedition to East Africa in 1856, Burton received a grant of £1,000 from the Royal Geographical Society and was given permission to take two years of leave from the British East India Company. However, for all his fame and undoubted brilliance as a linguist and a swordsman, Burton was not universally popular. He had been thrown out of Oxford University and had several enemies in the army; Richard Burton made enemies as easily as he made friends.

THE TEAM

Originally Burton had thought about taking a party of three or four men on the expedition but ultimately the only other European to join him was John Hanning Speke, a fellow officer from the East India Company. A few years earlier, Speke had been part of an expedition to Somaliland which had also been led by Burton. It had been an utter disaster: the British party had been attacked by local tribesmen, who were suspicious of their motives. The men were lucky to escape alive; Burton was severely wounded by a spear which entered one side of his face and exited on the other. Speke's wounds weren't quite so dramatic but they were equally grave. A year later Speke was invited to join the East African expedition at Burton's request, but the two men had little in common; whereas Burton was a skilled linguist with a keen eye for ethnography, Speke's main interest was in big game hunting. Burton was a worldly eccentric who was as interested in the sexual mores of the people he encountered as he was of their systems of trade. Speke was open-eyed but reserved; in many ways rather prudish. As the expedition progressed the tensions between the two men became more and more apparent.

EQUIPMENT

Anticipating that he might be gone for two years, Burton took a vast amount of equipment and supplies with him. A large part of this was intended to be traded for food and assistance and, more importantly, for paying 'hongo' to local chiefs. Hongo was either seen as a tax or a bribe; either way it was an absolute necessity. So too was a large supply of weapons and ammunition, for hunting and self-preservation. Burton and Speke took 400 different varieties of beads, rolls and rolls of cloth and drums of copper wire and their arsenal included 100 lbs (45 kilograms) of gun powder, 380 pounds (170 kilograms) of lead bullets and 20,000 copper caps. The two British explorers had revolvers and rifles; their guards carried much more primitive matchlock muskets. Miscellaneous items included a Union Jack, seven canisters of snuff and 2,000 fish hooks.

THE PLAN

The Ancient Egyptian historian Ptolemy had written that the source of the Nile lay in a range of mountains in the centre of Africa called the 'Mountains of the Moon'. No one was quite sure whether these were mythical or real. Burton's expedition was inspired by reports from Arab slave traders and European missionaries of a large lake which lay inland,

RICHARD BURTON'S CARAVAN
★ ★ ★

13 Baluch mercenaries on permanent escort	Said Bin Salim, the caravan manager
34 Baluch mercenaries on temporary escort	2 servants, Bombay and Mabruki
10 slaves hired as soldiers and interpreters	2 Goanese gun carriers
	39 porters
	5 riding asses
	30 asses employed as pack animals

to the east of Zanzibar. His goal was to chart this lake and try to ascertain whether it might be the source of the Nile. It was not going to be easy: the Frenchman M. Maizan, the last European to attempt to explore the region, had been hacked to pieces by members of the Kamba tribe; the culprits had never been found. Burton recognised that the very fact that he and Speke were travelling as explorers made them objects of suspicion; few people believed that anyone could be willing to suffer the pains of African travel for the sake of 'Science'. They assumed that there must have been an ulterior motive.

Burton and Speke left Kaole on the East African coast on 27 June 1857, after spending six months in Zanzibar equipping the expedition and making some preliminary forays into the nearby coast. They had a large caravan but there were far fewer porters than Burton had wished for; the caravan leader, Said Bin Salim, had estimated that they needed 170 men but he had only been able to recruit 39, because everyone was so frightened of travelling inland. Porters were happy to travel with Arab slave traders with whom they were familiar, but no one was keen on working for a *m'zungu*, a white man. In order to calm everyone's nerves, Burton consulted a local witch doctor who, after a suitable payment had been made, foretold that the expedition would be a huge success.

The caravan was unwieldy and difficult to control, a 'mob' as Burton described it. It moved slowly, averaging 2.5 miles per hour; on a good day Burton might spend five hours on the march but often it was less. Much of his time in the early stages was taken up with quelling incipient mutinies. The Baluch mercenaries had been attached to the expedition by the Sultan of Zanzibar, to their evident displeasure. They continually complained and on several occasions came close to mutiny. In one apocryphal story, Burton was said to have stabbed a mercenary after overhearing him plotting his murder. Said Bin Salim was weak and nervous and gave too much of their trading supplies at an early stage. The ten slaves brought along as servants and soldiers were anything but servile. They refused to carry any loads or be ordered around by the *jemadar*, sergeant, of the mercenaries. Burton was fluent in Arabic but Speke had to use an interpreter to talk to Said Bin Salim or any of the Arabs that they met on their journey. Neither man could speak any of the local African languages though at the end Burton claimed to have amassed a vocabulary of 1,500 words.

The other major problem for Burton and Speke was getting through the expedition alive. Their journey took them through a region where disease was rife and insects a perpetual irritation; smallpox and cholera were common and malaria was a constant presence. By the 1850s, quinine was starting to be used as a prophylactic against malaria but Burton relied on a quack concoction, 'Warburg's Drops', in which quinine was mixed with opium and bitter aloes. Though he slept under a net, Burton thought that it was simply a coincidence that attacks of 'fever' seemed to occur most frequently in areas where there were mosquitoes present. For long stretches of the expedition both men were so ill that they had to be carried on *tipoi*, hammocks attached to poles.

For Burton and Speke, just getting through the expedition was a triumph in itself. It is quite remarkable to read the accounts of all the pain that they suffered without seeming to be put off reaching their goal. In the most unpleasant incident, a beetle crawled into Speke's ear and proceeded to make it its home, its latrine and its burial place. Speke tried pouring hot butter into his ear to wash it out but to no avail; in a more desperate act he attempted to spear it with a penknife and ended up, in his own words, doing 'more harm than good'. The beetle ate a hole between Speke's ear and his nose, such that when he blew it, his ear whistled. Six months later, dismembered sections of the beetle began emerging in his earwax. Burton also suffered from mouth ulcers so severe that they 'rendered it necessary for me to live by suction', and recurrent attacks of 'fever'.

So how did Burton and Speke manage to get through it? In the first instance, they were ambitious and knew that if they found the source of the Nile they would be swathed in glory. Also though, you sense that their personal arrogance helped them to keep going. Both men felt superior to both the environment and the people around them; they weren't going to be put off by physical discomfort, however severe.

After five months, the caravan stopped at Taborah. Burton enjoyed the hospitality of the local Arab traders but Speke was frustrated by his inability to speak to them. Apart from Burton, the only other person that Speke could talk to was an African servant Bombay, who had learnt Hindustani in India. The Arabs confirmed to Burton that water lay ahead but they suggested that there were at least three different lakes. After several weeks resting at Taborah, Burton found it difficult to get the caravan going again; many porters had deserted and the remainder were not eager to move off. After much cajoling they left the town and on 13 February 1858, Burton and Speke reached a hill overlooking Lake Tanganyika. They were the first Europeans to set eyes on the world's longest lake but initially there was little celebration. Another bout of malaria had left Burton temporarily paralysed and Speke was suffering such severe ophthalmia that he could barely see. To add to his discomfort, on the way up to the brow of the hill, the ass that Speke was riding dropped dead from exhaustion.

Burton and Speke spent the next three months at the small settlement of Ujiji, recuperating and attempting to explore the lake. It wasn't easy

though; Lake Tanganyika is 400 miles (644 kilometres) long and 50 miles (80 kilometres) wide, its surface area greater than the whole of Belgium. Speke spent ten frustrating days trying to hire a *dhow*, a sailing boat, from a local Arab only to be told that he was welcome to have it, but unless he waited for several months he would have to find his own crew. Then Burton persuaded a local chief to take the two men on a journey by canoe to the River Rusizi, a large river connected to the north of the lake. Could it be the beginning of the Nile? It proved to be a frustrating journey; hostile tribes impeded their progress and when they got close to the Rusizi they were told that it flowed *into*, not out of, the lake. It was a big disappointment, but Burton still felt that on the whole the expedition had been a success. He had amassed enough material for two books and at least made a start on mapping the lake region.

After another month, they left Ujiji and headed back towards the coast. Again, they stopped at Taborah; whilst Burton recuperated from another bout of fever, Speke set off north to investigate another lake. He travelled quickly and on 3 August 1858, he set eyes on Lake Nyanza, which he immediately renamed 'Lake Victoria'. Speke spent three days camped next to the lake and then returned to Burton with the startling news that he had found the source of the Nile. Speke based his assertion on the simple fact that the lake that he had found was enormous and that, according to his measurements, it was 2,000 feet (600 metres) higher than the Nile when it reached Egypt. Speke didn't have time to explore the lake properly, but he had managed to talk to some local people through an interpreter. They told him that the lake was so big that no one knew its limits. Unsurprisingly, Burton was not utterly convinced by Speke: 'his convictions were strong, his reasons were weak'. Burton needed much more proof but he didn't want to go back with Speke to make further investigations. Instead he suggested that they should return to Britain to solicit funds for a second expedition. On the way back to the coast they agreed not to talk about Lake Victoria, lest they should have more arguments.

As a foretaste of things to come, Speke came down with another attack of fever, and in a fit of delirium poured out his resentment against Burton. Speke's illness worsened until he felt so close to death that he called for some paper to write a final letter to his family. However, after careful nursing by Burton and the servant Bombay, Speke recovered only for Burton to fall ill again.

They reached Zanzibar seven months later and travelled together to

Aden. At this point they separated; Burton stayed on to convalesce, while Speke headed back to Britain where he announced his great discovery. When Burton returned he was furious; Speke had stolen his thunder and persuaded the president of the Royal Geographical Society to fund another expedition to East Africa to confirm *his* findings. Though Burton was a much more credible explorer, he had enough enemies to ensure plenty of support for Speke.

The two men wrote a series of articles which downplayed each other's contribution to the first expedition and questioned each other's judgement. There was a suggestion that Burton might mount an expedition of his own to East Africa, but instead went on a long journey to Salt Lake City in America and was then appointed British Consul to West Africa. In the meantime, Speke returned to Africa with a friend, James Grant. This time he was able to make a much more detailed reconnaissance of Lake Victoria and was able to identify a large waterfall at its northern end as the source of Nile itself.

Speke returned to Britain to huge acclaim but he still lacked the cast-iron evidence that was needed to prove beyond doubt that he had found the source of the Nile. It turned out that Speke had not circumnavigated the whole of the lake and that at the point where he discovered the outlet from the lake, he was on his own, his companion Grant being too ill to travel. Speke had attempted to travel downstream from Lake Victoria to the coast as a definitive proof of his claim, but he had to miss out one section at the head of the river.

Speke's claim still hung in the balance in 1864 when he was due to meet Burton at the British Association in Bath. On the day before their debate, Speke was apparently very agitated and had been particularly uncomfortable sharing a podium with Burton. He visited an uncle who lived nearby and the two men went out on a hunting trip. As he was climbing over a wall, Speke's shotgun discharged itself leaving him mortally wounded. It was a very strange accident: Speke was an experienced hunter, so it seemed odd that he wouldn't have taken more care. Some conjectured that it might have been suicide, but the shotgun had such long barrels that it would have been virtually impossible for Speke to have pulled the trigger.

In his obituary, the *Times* came down heavily on Speke's side as the discoverer of the source of the Nile but it took the later travels of Baker, Livingstone and Stanley to definitively map out the route from Lake Victoria to the Nile Delta.

The sorry tale of Burton and Speke shows the dangers of the glory-hunting approach to exploration. If both men hadn't been so egotistical, they might have avoided all the controversy which damaged both their reputations. When Speke returned after his first journey to Lake Victoria, Burton was justified in his scepticism but not in the dismissive way in which he treated Speke. In his account of the expedition, he compared Speke's proofs to Shakespeare's Lucetta, the heroine of *Two Gentlemen of Verona*, who explained her liking of Proteus by saying 'I have no other reason but a woman's reason, I think him so because I think him so'. This was written after the event but the fact that, at the time, Speke had to agree not to discuss his 'discovery' during the return journey indicates just how competitive the two men were.

Both Burton and Speke received gold medals from the Royal Geographical Society for the African expeditions they led. They clearly possessed huge stamina and enormous will-power. Of the two men, Burton left the greater legacy, but for all the ethnographical details packed into the 43 volumes that he wrote about his expeditions, he was limited in his worldview. He showed far less sympathy and understanding of the Africans that he encountered than other British explorers like Livingstone or Stanley and his journals were both verbose and hampered by over-elaborate, self-congratulatory prose.

Speke was in some ways a more sympathetic character and was certainly more loyal and grateful to his African assistants, the 'Faithfuls', who accompanied him on his second expedition. His downfall was his impetuousness: if he had been more circumspect, if he hadn't been so eager for fame, he might have done things more carefully and staked his claim more modestly. The problem was that, rather than have a reasoned, scientific debate on what they had discovered, Speke and Burton flew into a very personalised debate which quickly degenerated into a slanging match. On the other hand, perhaps it is too harsh to criticise the two men for glory hunting; finding the source of the Nile had, as Burton noted, been the ambition of monarchs for thousands of years so the stakes were very high.

The Search for 90° South

One of Thor Heyerdahl's heroes was a fellow Norwegian, Roald Amundsen. Amundsen was one of the greatest polar explorers of the twentieth century: he was the first man to take a ship through the Northwest Passage and an airship across the North Pole. His most famous exploit, though, was his attempt on the South Pole in 1911. This brought him into direct competition with the British explorer Robert Falcon Scott who also made an attempt in the same year. Inevitably, the two expeditions were seen as a 'Race to the Pole'.

This race provides a fascinating opportunity to compare two very different approaches to polar exploration. Amundsen and Scott shared the same goal, but in other respects they were very different.

THE LEADERS

In his biography Amundsen wrote that, when young, he was inspired to become an explorer after reading about the Franklin expedition which was lost in the Arctic in the late 1850s. In particular, he said, he was inspired by

---

A BRIEF HISTORY OF THE ANTARCTIC
✶ ✶ ✶

1699	Edmond Halley leaves England on his ship *Pink Paramour* to search for the mysterious Terra Australis Incognita. He reaches 52°24'S, and sketches icebergs in the ship's log before retreating northwards
1772	On his second expedition, Cook crosses the Antarctic Circle and sails to within 890 miles (1,432 kilometres) of the Antarctic coast before being forced back by the pack ice
1821	The crew of the American sealing ship *Cecilia* become the first men to set foot on Antarctica
1899	Carsten Borchgrevink and two other men clamber the Ross Ice Shelf and sledge south to 78°50'S
1901	Robert Falcon Scott leads the British National Antarctic expedition. He and three other men reach 82°16'S on 30 December 1902
1908	Ernest Shackleton leads an expedition which reaches 88°23'S, a mere 97 nautical miles from the South Pole.
1911 (January)	Roald Amundsen and Robert Falcon Scott reach Antarctica and begin their 'race' to the South Pole

---

reading accounts of the suffering endured by the men on board Franklin's ships. Amundsen got his own first taste of expedition suffering in the late 1890s on board the *Belgica*, the Belgian ship which famously was held in the Antarctic ice for over a year. He learnt a lot from this experience; a few years later when he put an expedition together to travel the length of the Northwest Passage, everything went relatively smoothly and efficiently. After completing this historic journey Amundsen began to plan another expedition to the Arctic but when in September 1909 he heard that Peary and Cook were claiming to have reached the North Pole, he changed his plan and decided to head south, to the Antarctic. Initially though, he told no one but a few close confidants.

When in the same month Robert Falcon Scott announced that he was mounting an expedition to the South Pole, few people doubted that he would return in triumph. In 1901 Scott had led a major British expedition to Antarctica. There had been many problems but over all it was judged to be a success: Scott and a small party set a new record for going the furthest south and the scientists in his team carried out a lot of valuable work. Scott returned to Britain and then to the Royal Navy in 1904 but he regarded the Antarctic as unfinished business. He clearly felt that it was *his* territory and after his sometimes friend, sometimes rival, Ernest Shackleton *almost* reached the South Pole in January 1909, Scott was galvanised into action.

FINANCE

Both Scott and Amundsen had problems financing their Antarctic ambitions. Norway was a small country on the margin of Europe, and it was always difficult to find support for any expedition. Amundsen received a small grant from the government but he had to mortgage his house and buy a lot of his equipment and stores on credit. He did manage to borrow a very well-equipped ship though, the *Fram*, from his mentor, the great explorer Fridtjof Nansen. Scott's expedition was conceived on a larger scale, so his financial problems were more acute. The British government gave him a grant of £20,000 and he raised £10,000 in private donations but this was nothing compared to the £90,000 which had been secured for the first expedition. In 1900 Scott had been able to commission a purpose-built craft but this time he had to make do with the *Terra Nova*, a slightly run-down sealing ship. For both men, lack of money was a pressure they could have done without but neither of them was too worried about going into debt. They both assumed that when they returned from the Antarctic in triumph, they would be able to recoup the money through lecture tours and books.

THE TEAMS

Amundsen had a strong team: two expert dog drivers, a champion skier, two naval officers and a talented carpenter. None of them knew what they were getting into when they left Norway, but all agreed to follow Amundsen when he told them that he was heading for the South Pole, and not the Arctic. Amundsen's only potential problem was Hjalmar Johansen, a veteran Arctic explorer who had been with Nansen on his famous Arctic-drift expedition. Johansen was by far the most experienced hand on board the ship but he was a difficult character; he had a drink problem and at the age of 43 he seemed to feel that life had not treated him properly. Amundsen only agreed to take Johansen at Nansen's insistence; as he was borrowing Nansen's ship, he had no choice.

Scott's team was much larger. Whereas Amundsen landed on Antarctica with seven men, Scott had a shore party of 33. Five of them had been with Scott in 1901; apart from some civilian scientists, everyone came from the forces, principally the Royal Navy. Unlike Amundsen who was singularly focused on reaching the South Pole, Scott was always at pains to stress that his was a scientific expedition. On board were meteorologists, ornithologists, geologists, physicists and biologists who were eager to study everything from the breeding cycle of the Emperor King Penguin to the vagaries of Antarctic weather. They were all British, except for a Norwegian, Tryggve Gran, whom Scott hoped would teach his team how to ski, and two Russians who had been taken on to look after the dogs and the horses. Whereas Amundsen had a compact party of men who all had some experience of travelling across snow and ice, Scott's team was far less qualified. He had only one good dog driver and few of his men were comfortable on skis. Scott's diary is full of references to how keen and strong they all were, but they rarely backed up their enthusiasm with solid experience.

EQUIPMENT

Amundsen's methodical approach to clothing and equipment was one of his great strengths. Like Peary and earlier American explorers, he was a convert to the Eskimo style of travelling, using dog sledges for transport and wearing animal furs to keep the cold at bay. During his North West Passage expedition, he had lived with and closely observed the Netisilik

Eskimos of Alaska and had learned much from them. For the South Pole expedition, he ordered 100 huskies from the Danish Inspector of Greenland, telling him to spare no expense. Historically, British explorers had been less keen on dogs. When Ernest Shackleton returned from his attempt on the South Pole in 1909, he had written that neither dogs nor skis were of much use on the high Antarctic plateau. Instead he favoured ponies and man-hauling. Amundsen was sure that Shackleton was wrong and was determined to prove it. He felt that dogs were perfect for glacier travel because they were strong and light, and the fact that they were carnivorous meant that they could be fed on locally hunted meat. Amundsen was also a fervent advocate of skiing. All of his men were good, experienced skiers and Amundsen commissioned special extra-long skis for this expedition.

In spite of his experience as a naval commander, Scott didn't have Amundsen's logistical skills. When planning his expedition, Scott tended to cut everything very fine and not leave much room for error; in as unpredictable a region as Antarctica, this was a risky strategy. Scott didn't share Amundsen's obsession with equipment. Yes, he took it seriously, but he wasn't willing to put as much time or work into getting it right. Sleeping bags are an excellent example: whereas the Norwegians enjoyed warm, dry nights in their three-layer bags, the British explorers shivered in reindeer-skin bags that quickly filled up with ice. One of Scott's men, Apsley Cherry Garrard, described how the sleeping bag that he took on his famous winter journey weighed 18 pounds (8 kilograms) when he set off, and 45 pounds (20 kilograms) when he returned. There were some innovations on Scott's expedition; he took three tracked-vehicles, which could be seen as the forerunners of the modern Snow cats and tractors that are used all over the Arctic and Antarctic. Scott made a serious error in including the vehicles in his polar journey. Whilst they were fascinating for experimental purposes, they were just too new and unpredictable to be used on a dangerous journey.

Like Shackleton, Scott decided to use ponies as the expedition's main pack animals, rejecting the advice of the Norwegian and American explorers, who almost universally favoured huskies. Whereas Amundsen was planning to take some of his dogs all the way to the pole, Scott always intended that the bulk of the sledge-hauling would be done by the men themselves. The ponies would only be used for the first part of the expedition. Scott believed that there was something intrinsically noble about man-hauling and after a bad experience on his first expedition, he had no confidence in dog teams. The idea of cold-bloodedly calculating how many huskies to take and when to start feeding them to each other was anathema to Scott, just as the idea of forcing the men to haul their own sledges across the frozen wastes was anathema to Amundsen.

However, in order to hedge his bets, Scott did bring a team of 35 Siberian huskies as well as a small herd of ponies. This was typical of his slightly muddled thinking. Amundsen had two modes of transport: dogs and men. Scott had four: ponies, huskies, tractors and his favoured option, man-hauling. All of these were used in the expedition to the South Pole, making the whole thing that much more complicated.

THE PLAN

Amundsen was well aware that by heading south, he would be entering into direct competition with a British expedition, but he didn't pay much attention to Scott's territorial claims. In order to forestall some of the criticism, though, he decided to set up his base over 100 miles (161 kilometres) to the east of the British, on the Great Ice Barrier. This huge ice-shelf was almost 400 miles (644 kilometres) long and had walls up to 200 feet (60 metres) high; it stood between the Arctic Ocean and the continent itself. Once established on the Barrier, Amundsen planned to travel to the South Pole on as straight a path as possible, following a line of longitude. Unlike Scott's plan, this would mean finding a completely new route with all its attendant dangers and uncertainties.

Scott's plan was to head for the South Pole along the same route that Shackleton had used in 1908. He intended to set up his base in McMurdo Sound, an area that he knew well from his first expedition. His journey to the South Pole would take him over the Ice Barrier, up on to the Beardmore Glacier and then, finally, across the Antarctic Plateau to the South Pole. Scott's route had the advantage of familiarity: instead of striking out into the unknown like Amundsen, Scott would be passing through known territory. The disadvantage was that Scott's base was 60 nautical miles further north than the Norwegians', so his journey was significantly longer.

Scott's other big disadvantage was that when he left England, he knew nothing of Amundsen's intentions. He first heard the news via a tersely worded telegram that he received when his ship reached Australia: 'Beg leave to inform you Fram proceeding Antarctic. Amundsen.' Initially it wasn't clear exactly what the Norwegians were up to, but within a few

SOME DONATIONS TO
SCOTT'S LAST EXPEDITION
* * *

Eton College	A pony called Snippet
Westminster School	A pony called Blossom
The Liberal Club, Tyne and Wear	A pony called Jehu
Brighton and Hove Grammar School for Girls	A tent
Trafalgar House School, Winchester	A sleeping bag
King's College School, Wimbledon	A pony sledge
Copthorne School	A dog sledge
Tottenham Grammar	A man-hauling sledge
Royal Masonic School, Bushey	A motor sledge

months some of Scott's men would meet Amundsen in the Antarctic and report back to Scott that, indeed, the race was on.

THE RACE BEGINS

Amundsen arrived in Antarctica in January 1911 with 110 huskies (ten extra dogs having been born on the voyage down) and seven companions. They quickly unloaded their supplies and then moved inland to set up their camp, *Framheim*, two miles into the Barrier. Immediately they started hunting seals and before the winter set in, they amassed 60 tonnes of meat. Amundsen was a firm believer in eating as much fresh meat as possible in order to ward off scurvy; he had a lot of food from Norway but whenever possible both men and dogs ate fresh meat.

A month later, he and his men headed south on their first sledge journey. Their aim was to lay down a large depot of supplies that would be used on their attempt on the Pole. After the long sea journey, both the men and the dogs were out of condition and several huskies died. Overall though, this first foray on to the ice was a huge success. Amundsen managed to lay down three substantial supply depots, a vital boost for the race to come.

By contrast, bad weather and bad luck characterised Scott's expedition from the very beginning. Two ponies and one husky died even before his ship reached Antarctica. It took the *Terra Nova* three weeks to get through the pack ice, almost twice as long as Amundsen. When Scott finally started unloading, one of the three tractors sank through an ice floe and was never seen again. Scott's depot-laying journey was equally disastrous. Of the seven ponies he and his men took, six died. Several of the men had lucky escapes. Whereas Amundsen was able to lay down three food depots at 80°, 81° and 82° Scott didn't even manage to get to 80° south.

The sun went down in late April and for the next four months the two teams spent most of their time in their respective huts. Amundsen's men used the time to systematically overhaul almost every single item of their equipment. The principle was simple: everything had to be as light as possible, as reliable as possible and as efficient as possible. The sledges were rebuilt to less than a third of their original weight. They unstitched their boots and redesigned them so that they were warmer and more comfortable. They even planed down their supply boxes so that they were 30 per cent lighter and then, to cap it off, they worked out a way of loading them on to the sledge so that the food could be taken out without having to unpack everything else. This stage of the expedition showed Amundsen at his analytical best: he knew that the journey would be a real test for everyone so he worked hard to save every single ounce of weight.

During the winter months, Scott's men spent a lot of their time on scientific work. One group made a daring trip to a penguin colony; it was a brutal journey, made in the dark, freezing cold winter which left them drained and exhausted, hardly fit for the rigours to come. Many of Scott's men did work on their equipment, but they just didn't have the same thorough approach as the Norwegians. In the middle of May, Scott revealed his complex plan: it involved various teams of dogs, ponies, tractors and men setting off at different times across the ice, then gradually peeling away until a small group made the final attempt on the Pole itself. He was hopeful of success, but Scott, like Amundsen, was worried about the competition.

Amundsen knew that it would be a personal disaster if he didn't reach the Pole first. He would be denounced as both a cad, for failing to announce his expedition, and a failure, if he didn't beat Scott. When the sun came up for the first time at the end of August, Amundsen was desperate to start but it was just too cold. Fifteen days later he and his seven companions left their hut, with four sledges, 86 dogs and high hopes. It was a big mistake.

In his eagerness to get to the Pole, Amundsen had jumped the gun. At first the temperatures were reasonable and the going was good, but then the cold weather returned with a vengeance. It was too cold for the dogs to perform well and two of the men developed frost-bite. As soon as they reached their first depot at 80°, they turned around and beat a hasty and chaotic retreat.

Back at the hut, a furious row erupted between Hjalmar Johansen and Amundsen; the polar veteran accused Amundsen of panicking and questioned his decision to set off in the first place. This was a very public challenge, which called into question Amundsen's leadership of the expedition. He had to respond decisively. After talking to the men individually and extracting pledges of loyalty, he made a dramatic change of plan: instead of taking all eight men to the South Pole, he would split the party, leaving Johansen and two others to explore King Edward VII Island whilst he took four men to the South Pole. This was a key moment: Amundsen had swiftly and firmly reasserted his leadership and, by reducing the size of the party, he had given himself a greater margin of error. Earlier in the year, they had cached enough food for a party of eight out on the Barrier but now it would be shared among just five men. Amundsen never forgave Johansen, but inadvertently his mutiny had increased their chances of success.

A month later, five Norwegians set off again with four sledges and 52 dogs. Amundsen's schedule required him to travel one degree of latitude every four days, but he built in a margin of safety for bad weather. Everyone was very keen but Amundsen was always reluctant to force the pace, not wishing to tire the dogs. The weather was mixed: sometimes there were blue skies but sometimes the visibility was appalling. Most of the time the Norwegians travelled on skis and whenever they could they all roped up to the sledges to be towed along.

Scott began his attempt on the Pole almost two weeks later than Amundsen because his ponies couldn't cope with extremely cold weather. The expedition began badly: the tractors broke down repeatedly until they finally gave up altogether barely 60 miles out of their base. Their failure delayed the schedule and had the knock-on effect of increasing the amount of supplies that had to be man-hauled. The ponies found the going very difficult but the dog team performed much better than he had expected. It was too late now though for Scott to change his mind; he kept the dogs on for longer than he anticipated but they were sent back to base after six weeks. In early December Scott and his men arrived at the Beardmore Glacier, their highway to the Antarctic Plateau. The ponies that had survived the first stage of the journey were slaughtered; from now on, everything would have to be man-hauled.

Amundsen's team travelled much more quickly; by early December they had reached the Antarctic Plateau itself. The dogs performed incredibly well; on one day they covered 17 miles and made an ascent of 5,750 feet (1,725 metres). Nevertheless, at the appropriately named Butcher's Camp, 24 of the huskies were killed; each driver had to slaughter the requisite number of dogs from his own team. No one relished killing their animals but they all knew that it was a necessary part of the whole plan, and that the fresh meat would help protect them against scurvy.

After being held back for five days by a bad storm, Amundsen's men broke camp. For Baajland, the champion skier, the challenge of going ahead to find a route was one to relish. Amundsen's leadership was instinctive and sophisticated: having decisively quelled Johansen's mutiny, he made sure to delegate responsibilities. He did his fair share of the camp chores and frequently took advice from the others on the best route to take. And it wasn't easy: the names they gave to the terrain they crossed tell their own story – the Devil's Glacier, the Gates of Hell, the Devil's Ballroom.

Meanwhile, as they moved up the Beardmore Glacier, Scott's men came to realise just how hard it was to man-haul their sledges over broken terrain. Modern-day apologists for Scott, such as Sir Ranulph Fiennes, point out that in recent years there have been several successful Arctic and Antarctic expeditions based on man-hauling, but they are not comparing like with like. Today's man-haulers have better sledges, better clothing, better food and better communications. Scott's team were all very brave and very strong, but man-hauling was exhausting, time-consuming work which made a dangerous journey even more risky.

On 14 December 1911, whilst Scott's men were still struggling up the Beardmore Glacier, Amundsen arrived at the South Pole. Amundsen searched around gingerly, but he could see no sign of his British rivals. He and his men took a series of sextant readings and moved position twice, until they reached what Amundsen computed to be the South Pole itself. Then he sent everyone out to make a circuit of 12.5 miles (20 kilometres) in radius so that no one could accuse them of missing their goal. Finally, after leaving a small tent with a message for Scott and a letter to the King of Norway, they were off. Scott, ironically, would be the one to carry the proof that Amundsen had succeeded.

A month later, Scott saw the first sign of Amundsen's expedition, a black flag in the snow. It was a bitter, bitter blow. In his diary, Scott put a brave face on it but his hurt showed through: he had been 'forestalled', he had said goodbye to all his 'daydreams' and now all they faced was 800 miles (1,287 kilometres) of drudgery. Amundsen returned from the Pole flushed with victory but nervously anticipating all the glory to come. Scott had nothing to look forward to except the end of the journey. He had by now reduced his team to a party of five men. All of them knew that it would be a long, difficult journey home.

Amundsen's return, like the journey out, was controlled and relatively uneventful. There were some difficult days when he lost his bearings and wasn't sure where their food depots were, but this crisis didn't last long and they were quickly back on track. When they finally left the plateau and moved down on to the Barrier, Amundsen's men had so much food that it felt like 'the fleshpots of Egypt'. They arrived back at their hut a full ten days ahead of schedule. The *Fram* was lying nearby and soon they were all aboard. A month later Amundsen cabled the world from Tasmania: 'Pole attained 14th–17th December 1911. All well.'

By early March, Scott's men were far from well. Their return journey initially went smoothly but after 20 days, the first member of Scott's party, seaman Evans, died suddenly and mysteriously. Bill Wilson, one of the expedition doctors, didn't know what to make of it. Was it brain damage caused by a fall? Or was it the onset of scurvy? Or pure exhaustion? No one knew, or had time to ask. The race to the Pole had now been transformed into a race for survival.

A month later, on 16 March, the second man died, Captain Oates. His feet were severely frost-bitten and he was tormented by an old wound picked up in the Boer War. According to Scott's diaries, one morning Oates announced, 'I am just going outside and may be some time,' before leaving the tent, never to return. The three remaining men struggled onwards and managed to reach the Barrier. There they found their caches of food but, to their horror, there was not as much oil as they expected. Their stove ran on oil; without it they could neither cook nor melt water. To make things worse, the weather was much colder than they had expected so they needed the fuel even more. It was later discovered that a lot of the oil had simply leaked out of the cans left in the depot. Whereas Amundsen had so much left at the end that he cached a 4-gallon (17-litre) can of paraffin in a cairn on the edge of the Ice Barrier, Scott was running desperately low at the very point when he needed it most.

As they tramped across the Barrier, Scott vainly hoped that some of his team members might come out to help. He had left somewhat contradictory orders: in the first instance, he had asked Cecil Meares, the expedition's best dog driver, to come back at the end with his huskies to help the polar party in the final stages of the return journey. However, he also added that no one should take any risks with the dogs, in case they were needed for a third attempt on the Pole. Meares was not available to come out to meet Scott so the task fell to Apsley Cherry Garrard, who had no experience of dogs whatsoever. Cherry Garrard took Demetri Gerov with him but the Russian dog handler soon fell ill. Cherry Garrard stationed himself at the main supply depot on the Barrier, but he was a poor navigator and afraid that if he went any further he might miss the return party. He didn't have enough dog meat to stay for long and, mindful of Scott's warning not to jeopardise the huskies, Cherry Garrard dropped off some supplies, and left after six days.

Meanwhile, Scott and the others were less than 20 miles (32 kilometres) away, suffering terribly. At the very point when they were at their lowest ebb, they were hit by even lower temperatures and vicious storms. The three survivors spent their last week marooned in their tent hoping against hope that the weather would improve. They wrote poignant letters to their loved ones as the life trickled out of them. Eight months later they were found lying next to each other in their tent, frozen solid.

Next to the bodies, there were diaries and letters. Scott had written a final message to the public. It climaxed with an impassioned plea to the nation, to look after the team's 'people' but much of the text was aimed at defending his reputation. The unexpectedly bad weather, he wrote, had been their downfall and 'every detail' of their clothing, equipment and the laying down of depots, had worked perfectly. Looking at the record it is impossible to agree with this; there were just too many errors.

In recent years, Scott's reputation has waxed and waned. The polar historian Roland Huntford damned him as an incompetent in his best-selling book on the race for the South Pole. More recently the modern British explorer Sir Ranulph Fiennes launched a biographical defence of Scott, reclaiming him as a national hero. He pointed out that Amundsen too had made mistakes and that, based on his own previous expeditions, man-hauling wasn't such a bad idea. Fiennes was backed up by an American meteorologist, Susan Solomon, who had gone back through the records and discovered that Scott had, in fact, encountered a freakishly cold winter. Maybe the truth lies somewhere between the disparagement of Huntford and the adulation of Fiennes. Huntford had clinically exposed all Scott's flaws and memorably described him as a 'heroic bungler', but this is too harsh a judgment. Perhaps a fairer assessment is that he was a heroic 'amateur'. Scott was passionate about the Antarctic and fascinated by all the opportunities it offered for scientific research. He was interested in everything – meteorology, photography, geology – *and* wanted desperately to get to the South Pole. However, Scott under-estimated just how difficult a journey it was going to be. Later, Amundsen was criticised by some for being 'ruthlessly efficient' and 'professional' but it was precisely this approach that got him and his men to the South Pole and back – alive.

Amundsen did have faults and there was something uncomfortable about the way the expedition started, but once he got to the Antarctic, he was an excellent leader. It is easy sometimes to see leadership principally in terms of the personal relationships, but whilst it is the case that in-spiring teamwork, boosting morale and maintaining discipline are all important, leaders also have to make the *right* decisions, *most* of the time. Amundsen chose the best base, the best means of transport, the best clothing, the best equipment; he had the best plan and when he took risks he insulated himself with a decent margin of error. If Scott had had some of Amundsen's ruthless efficiency, he might have lived to tell the tale.

A SHORT HISTORY OF AUSTRALIA
✷ ✷ ✷

40–80,000 BC	'Aborigines' arrive in Australia
1503–16	Various Dutch, Portuguese and French explorers land on Australia
1642	Dutch explorer Abel Tasman lands on Tasmania and claims it for Holland
1699	Pirate William Dampier explores the north west coast and brings back to London reports of large hopping creatures
1770	Captain Cook charts east coast of Australia and claims the continent for the British Crown
1788	Following the suggestion of the botanist Joseph Banks, Australia is turned into a penal colony. The first 750 convicts arrive at Botany Bay and then settle at Sydney
1792 and 1822	Smallpox decimates the Aboriginal population
1815–40	Free immigration rises; the number of convicts declines
1850s	Australian Gold Rush brings mass immigration
1860	Robert O'Hara Burke is chosen to lead an expedition whose aim is to cross Australia from south to north

I have only one ambition, which is to do some deed before I die, that shall entitle me to have my name honourably inscribed on the page of history. If I succeed in that I care not what death, or when I die.

ROBERT O'HARA BURKE

When the first British settlers arrived in Australia, they built their towns and cities along the coast. The interior was literally a blank on the map: some thought that there might be a huge inland sea at the centre of the continent but no one really knew. The first expeditions to go further inland had mixed results: some ended in death and disaster, others were more successful, but no one managed to go all the way from the south coast to the north. Then in the late 1850s a race started between two of Australia's most prosperous states, Victoria and South Australia, to see who could make the first transcontinental journey. Victoria's attempt was led by a flamboyant Irish officer, Robert O'Hara Burke; he was an odd choice, more famous locally for his uncanny ability to get lost than for any skills as an outdoorsman.

The story of what followed is one of the great exploration tales, full of eccentric characters and strange coincidences. Burke was a very brave man with incredible stamina, but he was entirely unsuited to leading such an expedition. The lesson of this expedition may be obvious but it is amazing how many times it is forgotten: that leadership matters, and no matter how much money is thrown behind an expedition, if the correct leader is not chosen at the beginning then disaster frequently ensues.

THE LEADER

The Melbourne expedition committee dithered about choosing a leader. The simple fact was that the man most qualified to lead the expedition, John McDouall Stuart, was based in their rival state, South Australia. They had to look elsewhere. The committee was too class conscious to put a farmer or a bushman in charge; social position seemed as important to them as skill and experience. Like a lot of committees they were riven by factions and rivalries, so when good candidates did appear there was usually someone who had an objection to them. At one stage they even advertised in the newspapers for 'gentlemen desirous of offering their services for the leadership of the forthcoming expedition'. Eventually Robert O'Hara Burke was chosen; he gained his position more through cronyism than by dint of any real qualifications for the job. He was a rakish, rootless Irish gentleman who had been a soldier in the Austrian Army and a mounted policeman in Dublin before coming to Australia in 1853. He had no experience of exploration and had no instinctive bush sense, but he was charming and energetic and had the knack of impressing the right people. His appointment prompted controversy in the press but he and the other committee members rode out the storm.

FINANCE

The expedition was put together by a committee of the 'great and the good' of Melbourne; they saw it as both a way of asserting themselves over the other Australian states and potentially as the first stage of a land grab. After a gold strike in the 1850s, Melbourne had boomed and now the city fathers were looking for an outlet for their cash and their ambition. If they could win the race to the unclaimed north coast, they hoped that it might enable them to set up a new colony. Initially local enthusiasm wasn't matched with cash, but eventually the exploration committee persuaded the State to put up £6,000 for the expedition. This money was badly managed throughout the expedition; as Burke journeyed north, he left a string of bad cheques and a lot of ill feeling behind him because the committee didn't put enough money into his account. The money soon ran out and they had to procure another £6,000 grant from the State government. When many years later the final accounts were drawn up it was discovered that the whole expedition had cost £57,840, more than five times the original budget and several millions in today's money.

THE TEAM

Seven hundred people applied to join the expedition, but the same cronyism that had characterised the appointment of the leader influenced the choice of the team. Most of the men chosen had some connection to the committee. They were a mixed bag including five Irishmen, three Englishmen, three Germans, one American and four Indian sepoys to look after the camels. Few of them had any experience of travelling in the Australian bush. Second-in-command was George Landells, an Englishman whom the committee had employed to procure and transport 25 camels from India. He didn't get on with Burke and their mutual hostility caused a lot of problems in the early stages of the expedition. The expedition surveyor was William John Wills, a 26-year-old Englishman who had come to Australia during the gold rush. He had worked as a shepherd and a surveyor and had a passion for the Australian outback; unlike Burke, he was famous for his strong sense of direction. Though Burke only ever thought about getting to the north coast, the committee wanted this to be a scientific expedition so they employed a botanist and an artist to sketch the terrain along the route. Unsurprisingly, this caused some friction, and Burke treated the scientists badly. The expedition foreman was Charles Ferguson, an American horse-trader with a murky past. He would also cause problems for Burke.

EQUIPMENT

Burke and the expedition committee purchased equipment without any real sense of what was needed. The expedition was equipped with dandruff brushes, enema syringes, a bathtub and a formal dining table. Other gimmicks included Chinese gongs and rockets to be used if anyone got lost, an amphibious wagon and a special hospital stretcher that could be fitted to one of the camels. Most of these items were never used. It was smart, however, to take a herd of camels with them. This was the first time that they had been used on this kind of scale in Australia. Unfortunately, the camels became a source of friction between George Landells and Burke; Landells wanted them to be treated more gently, on the first stage of the expedition, but Burke would hear none of it.

STAGE ONE: TAKING 20 TONNES OF SUPPLIES
ON A 466-MILE (750-KM) JOURNEY WHICH
COULD HAVE BEEN DONE BY BOAT

On 20 August 1860 Burke left Melbourne with a cavalcade of 19 men, 23 horses, 27 camels (additional animals having been procured from a local circus) and six wagons. It was the beginning of a 3,107-mile (5,000-kilometre) journey to the north coast and back, which was expected to take up to two years. Their departure had been delayed by several weeks because of the late arrival of the camels; this had serious consequences for the schedule. Burke had to choose between delaying his journey or travelling during the hottest period of the year. He chose the latter. At first though, they had to contend with freezing nights and damp, muddy days which slowed the wagons down to a snail's pace. Their camels found the going very hard and were soon sick; their only success was in terrifying all the cattle and horses which they met on the way.

It was not a happy team: the going was difficult and Burke seemed oblivious to everyone else's problems. For the first few hundred kilometres they were travelling through farmland; Burke tended to ride ahead of the men during the day and often slept in local farmhouses at night if he could find someone to take him in. He insisted on a fast pace and seemed to have no particular interest in anyone's welfare. After three weeks, he had lost five men and taken on four replacements. The absurd thing about the whole first stage of the journey was that it could all have been done by boat along the Darling River. Burke turned down the idea because of two typically political reasons: first, he thought his supplies might be nobbled if they passed through the port of Adelaide, the capital of his South Australian rivals; and second he was reluctant to entrust himself to Captain Francis Cadell, who offered his steam boat for free. Cadell was associated with a rival faction on the Melbourne expedition committee and Burke didn't trust him. It was clear to everyone, though, that Burke's wagons were seriously overloaded and just 249 miles (400 kilometres) into the journey Burke auctioned off the first batch of his supplies. Some of the material was superfluous but Burke showed how little he understood about exploration by disposing of the supplies of lime juice that had been taken along to help guard against scurvy.

It didn't take long for the arguments to start. Burke quarrelled with his foreman, his scientists and his deputy, George Landells. He demanded that Landells' camels should carry heavy loads, to take the burden off the wagons, but Landells wasn't happy. Back in Melbourne, the committee, albeit ambiguously, had told him that he was in sole charge of the camels and he sensibly argued that their loads should be kept as light as possible until they reached the true desert, at which point they would take over as the main pack animals. Burke couldn't be persuaded.

After six weeks of hard slog they arrived at the Darling River. Burke impetuously decided on a change of plan: they would abandon the wagons and instead load eight tonnes of supplies on to a steamboat which would take them upriver to Menindee, the last town before the desert. Whilst waiting for the boat there were more furious rows and more moments of high farce. The camels wandered off into the scrub and the men spent two fruitless days searching for them, frequently getting lost themselves. Eventually they had to hire an Aboriginal tracker to find the animals. The tension between Burke and Landells came to a head when Burke sacked his deputy and shortly afterwards challenged him to a duel. It never happened but Landells headed back to Melbourne licking his metaphorical wounds.

By the time Burke and his remaining men reached Menindee it was the beginning of the summer. Several people advised Burke that he should wait the hot season out but he would have none of it. He was worried that his rival from Adelaide, John McDouall Stuart, would beat him to the coast, even though he heard news that Stuart had failed on his latest attempt. Once again Burke decided on a change of plan: he split the party in two and proceeded onwards, taking most of the horses and camels. He told the rear party that he would send for them soon but he left them demoralised and confused. More wisely he hired two Aboriginal guides and a local farmer, William Wright, to help him on the next stage.

Burke had stumbled his way through the first stage of the expedition without any serious mishap but the whole story was one of waste. By the time they reached Menindee, Burke had dispensed with several tonnes of supplies and was way behind schedule. He could have reached the town far more quickly if he had done the journey by boat. On the positive side, the expedition surveyor, William Wills, was proving himself to be a very able part of the team, but as a team leader Burke tended to alienate rather than inspire.

STAGE TWO: 10 MEN,
16 HORSES AND 19 CAMELS

Burke's smaller party moved much more quickly and over the next two weeks they made good progress. Burke was so impressed by Wright and his Aboriginal guides, that he decided to appoint him as his third-in-command and sent him back to Menindee, with orders to pick up men and the stores that had been left behind. Meanwhile, Burke and the others pressed on to the fabled Cooper's Creek, a vast river basin containing a network of lakes, rivers and pools. It was teeming with wildlife and a focal point for local Aboriginal tribes.

They were still 932 miles (1,500 kilometres) from the end of their journey though, and as they started to make exploratory forays into the surrounding scrub, they found that the landscape was much less forgiving and that water was much harder to find. As they waited for Wright to bring up the other men, their first camp at Cooper's Creek was overrun with thousands of rats from the desert, forcing them to move on. The flies and mosquitoes gave them no rest and Burke was, as ever, literally itching to get on with the journey.

Instead of waiting for William Wright and the others, or sending someone back to find out what was going on, after a month Burke elected to split his team yet again and take a still smaller party on a final dash to the coast. They would take the minimum of food and supplies. Burke told the men that he left behind to wait for three months; after that they should try to hold out but, if they had any problems, they were under no obligation to stay.

It is hard to imagine Burke acting differently, but anyone with experience of the bush might have paused. They were about to travel through the desert at the height of the Australian summer. They had no maps, no Aboriginal guides and they didn't even have any tents. Burke decided that they would be too heavy to carry so they left them behind.

STAGE THREE: FOUR MEN,
SIX CAMELS AND ONE HORSE

Initially things went well for Burke and his slimmed-down party; with fewer hands on board there were fewer arguments. They managed to average 16 miles (25 kilometres) a day and were usually lucky enough to find some kind of pasture for their camels. On the route north they frequently encountered Aboriginal tribespeople who offered them presents of fish; the Aboriginals were awed and frightened by the camels, otherwise they might have been more aggressive to these intruders in their territory. It was a gruelling march through a burning hot desert but Burke continued to insist on a relentless pace. They were lucky with water though and managed to find a branch of the Diamantina River. It was connected to another river system which ran all the way to the north coast but first they had to find a way across the difficult ground of the Selwyn mountain range, a stage which, Burke noted in his diary, left the camels 'sweating with fear'.

After 45 days they were still about 124 miles (200 kilometres) from the coast and had used up more than half of their stores; Burke, however, decided to charge on. Doing so meant that they had virtually no chance of returning to Cooper's Creek within three months and therefore no guarantee that the others would still be there when they returned. Again it's hard to see how he could have behaved otherwise but it was a fatal mistake. Knowing when to turn back is one of the hardest things for any explorer. Burke was now cutting his margin extremely fine; the desire for glory led him on but the risks were stacking up.

The landscape changed again as the party moved out of the desert into the damp tropical zone that fringed the coast. Their sufferings only increased; the desert had been bad enough, but at least they could sweat off the heat. Now the perspiration rolled off them and left them damp and hot: it was just too humid for their sweat to evaporate and cool them down. One camel wandered down into a creek and was unable to get out again. The men were too weak to rescue it, so they just redistributed their meagre loads and stumbled on. For the last time, Burke decided to split the party;

he and Wills carried on with the horse in tow whilst the other two, Charles Gray and John King, stayed put with the camels. Burke was desperate to end the journey at the ocean itself but a huge mangrove swamp blocked their way. They turned back roughly 12 miles (20 kilometres) from the coast, disappointed but satisfied that they had made the first crossing of Australia.

They had won the prize, but could Burke and Wills get back to claim it? They rejoined their two companions and began the long trudge back. The outward leg of their journey from Cooper's Creek had taken longer than Burke expected; it would be a race to get back to the others, but they had to try. The monsoon arrived and the ground turned to mud. Their camels were weak and the men were now marching on reduced rations because they had eaten too much of their supplies. They had guns and there was plenty of wildlife around them, but they were simply too tired to hunt; gradually the life ebbed out of them. In order to reduce the loads on the camels still further, Burke abandoned more equipment and more supplies, reducing their rations just when they needed them the most. Charles Gray, the largest man in the party, complained the most and one day they caught him stealing from what remained of the expedition's flour. Burke was furious; according to Wills' diary, he gave Gray a 'good thrashing'. Could there ever have been a more absurd waste of energy?

The men left the tropical coast behind and re-entered the desert; they were lashed by dust storms and tormented by hunger. Even after eating one of their camels and their ailing horse, they were still ravenous. The horsemeat was tender but there was barely a sliver of fat on it. Four months after they had left Cooper's Creek, and three weeks after his thrashing, Charles Gray the hungry sailor died. The others buried him and then, after finally abandoning almost all their remaining equipment and supplies, they trekked on. Their only hope was that the second half of their party would still be at Cooper's Creek but when they arrived, on 21 April 1861, they found that their old camp had been abandoned. There was a message carved on a coolabah tree: 'Dig, under, three ft NW.' There they found a trunk with a message inside telling them that the other men had left Cooper's Creek on that very day, having waited an extra five weeks for their return!

Meanwhile ... the men who had been left at Cooper's Creek were making their way wearily back to civilisation, assuming that Burke was dead. The long hoped-for relief party had never arrived because William Wright, the farmer taken on as third-in-command, didn't leave the outpost at Menindee for three months. Wright had been waiting for the expedition committee back in Melbourne to approve his appointment and give him permission to buy horses to carry the supplies out to Cooper's Creek, but they had not replied. When they finally authorised him to lead the relief party it was the middle of January and he was long overdue.

On the way out, Wright's party encountered the men returning from Cooper's Creek in the middle of the desert. They were overwhelmed to meet each other but no one thought that Burke and Wills could still be alive. Just to make sure, Wright and one other man rode on to Cooper's Creek. They stayed for just 15 minutes; seeing no sign of Burke and Wills, they turned around and headed back.

Two weeks earlier, Burke and the others had set off for a remote Australian police outpost at Mount Hopeless, 155 miles (250 kilometres) away. It was another fatal mistake. In theory it was closer than their camp at Menindee, but they were in no fit condition to make the journey and by heading west, away from Menindee, they made it virtually impossible for anyone else to find them. Wills hadn't wanted to go; he thought it would have been better to have stayed at Cooper's Creek or headed back along their outward trail but Burke insisted. Sinking a final nail in the coffin, they departed from Cooper's Creek, without leaving any visible notes and barely any indication at all that they had been there! This was an expedition whose whole story could be written in exclamation marks.

After two futile weeks tramping west, Wills realised that he didn't have the strength to continue. Instead, he turned around and headed back to Cooper's Creek to bury his journals next to the coolabah tree; thanks to this gesture at least the story of the expedition would be known. Burke soldiered on until he and the third man, John King, ground to a halt. Wills belatedly realised that their only hope of salvation lay in cultivating the local Aboriginal tribes, but when he rejoined the others, he found that Burke had just had a serious confrontation with the Aborigines and then, to top things off, had accidentally set his remaining possessions on fire. Over the next few weeks the men slowly starved to death; they didn't have the strength to hunt the birds and waterfowl around them and had given away or lost most of their fishing equipment. Eventually Wills felt so weak that he wrote a final letter to his father and gave it to the other two men, telling them that if they were to survive, they had to find the Aborigines and make friends with them. Burke died a few days after Wills. John King, a 23-year-old former soldier who had only joined the expedition to assist the sacked George Landells, was the sole survivor.

Two months later, on 15 September 1861, one year and 25 days after Burke's party set off from Melbourne, a rescue party arrived at Cooper's Creek and found John King. His clothes were in tatters but thanks to the Aborigines, he was still alive. When the news got back to Melbourne, Burke and Wills were lauded as true heroes, 'dead' proof of Victoria's ascendancy over the other southern states. Before long though, the arguments began. Why had so many men died? Some newspapers turned on the exploration committee, others criticised Burke himself. Finally, after overwhelming pressure, a public enquiry was held. There it was acknowledged that Burke had made some mistakes but the brunt of the criticism fell upon William Wright, the farmer and bushman who, according to his critics, should have brought a relief party up to Cooper's Creek at a much earlier date. He argued that it was the committee's fault for taking so long to confirm his appointment but no one heeded him and the committee was largely exonerated.

Today Burke is no longer the great hero. He was brave, but he was foolish, reckless and disorganised; he just shouldn't have been made the leader. His one really good piece of judgement was to make Wills his second-in-command, once he'd fired George Landells. Wills was the only one who actually understood the importance of learning from the local people, but by the time he realised this it was too late. The expedition is also a good example of the perils of organising committees: throughout

the whole affair the Exploration Committee was lazy and incompetent. The committee chose the wrong leader and failed to keep in touch with the expedition. All of the unhappy coincidences at the end make the expedition seem as if it was cursed with bad luck, but in this disaster poor judgement and bad management played a crucial role. Burke won the race to cross the continent but it was a hollow victory; he was an incompetent leader backed by an incompetent committee. It was a miracle more men didn't die.

EVEREST, THE CONTINUING SAGA
✱ ✱ ✱

First ascent by a woman:
16 May 1975, Junko Tabei

First ascent without oxygen:
May 1978, Reinhold Messner and Peter Habeler

First solo ascent:
August 1980, Reinhold Messner

First descent in a paraglider:
1988, Jean-Marc Boivin; it took just 11 minutes to reach the ground

First ski descent:
July 2000, Davo Karnicar

Estimated number of corpses on Everest:
120

In the late spring of 1996 Everest was once again front-page news when in a few short days 12 climbers died trying to reach the summit. Five of the victims came from two commercial expeditions, which made their attempts on 10 May. A year later Jon Krakauer's inside account of that fateful day, *Into Thin Air*, was published to great acclaim; it soon became one of the biggest-selling adventure books of all time. Since then a lot more has been written about the 1996 disaster, but there are still many unexplained events and unanswerable 'what ifs'. In one sense the deaths

were yet another reminder of the ever-present dangers of high-altitude mountaineering but from another point of view the 1996 tragedy was an inevitable result of the commercialisation of the world's highest mountain.

This is not a story of a failure of leadership, or of a breakdown in teamwork; rather it is a story about a culture change in mountaineering.

AFTER HILLARY AND TENZING

For most of the twentieth century, Everest was the preserve of the world's elite climbers. Initially most expeditions were from one country, but during the 1970s and 1980s a new style of international expedition emerged, uniting top climbers from all over the globe. Most of these were funded by business sponsorship or by individual members putting up their own money. Then something new happened: in the early 1990s commercial operators arrived on the scene offering places on their expeditions for a fee. No longer did you have to be an elite climber to make an attempt on Everest. Himalayan mountaineering developed into a growth industry. Governments charged expensive 'peak fees', Nepalese Sherpas hired themselves out as high-altitude porters and Western mountaineers found that they could make a small living by helping their predominantly rich clients to realise their dreams. Some elder statesmen spoke out against what they saw as the commercialisation and desecration of Everest, but no one took them seriously. After all, mountaineering in the Alps had always been a commercial activity, so this was nothing new. The big difference though was that in the Himalayas there were many more 'objective' dangers: in addition to the avalanches, crevasses and storms found in the rest of the world's mountains, the Himalayas offered climbers the enormous challenge of climbing at high altitude.

THE TEAMS

In 1996 the two biggest commercial outfits on Everest were Adventure Consultants, a New Zealand-based company led by Rob Hall, and Mountain Madness, an American company founded by Scott Fischer. This was Adventure Consultants' fourth commercial expedition to Everest. Rob Hall was a very experienced climber who had previously guided 39 climbers to the top. In 1996 he had eight commercial clients including Jon Krakauer, a mountaineering author sent along by an American magazine to write a piece on guided expeditions to Everest. Scott Fischer was a charismatic and successful American climber who had made a name for himself by leading a team of business executives up Mt Kilimanjaro, in a highly publicised sponsorship climb. He had climbed Everest in 1994 but this was the first time that he had run a commercial expedition on the mountain. He, too, had eight clients for his Mountain Madness expedition.

The clients varied widely in experience. Because the cost of going to Everest was so great, their average age was higher than on most expeditions. Adventure Consultants' official price was $65 000 per person and that meant that you had to be very rich, or very persuasive, to take part. Clients tended to be successful, highly motivated professionals; on Rob Hall's team there were three doctors, a lawyer and a Japanese business-woman. Most of them were reasonably experienced and one man was making his fourth attempt on Everest. However, as Rob Hall continually emphasised, none of them was capable of getting to the top without the help of *his* guides and *his* Sherpas. Scott Fischer's team was slightly younger and more eclectic; there were two ski-patrollers from Colorado, a construction worker and a New York socialite who, on a previous attempt, had taken her nine-year-old son and a nanny with her to base camp.

In this context, the word *team* is something of a misnomer. In 1953 John Hunt had led a British and Commonwealth team to the summit of

Everest; they were all strong climbers who genuinely worked towards a common goal. All of them were ambitious but they realised from the beginning that only a few of them would ever get the chance to make an attempt on the summit, however, for that attempt to take place the others would have to put in an enormous effort to prepare the way. In 1953 the Sherpas performed heroically as high-altitude porters but, apart from Tenzing, few of them had any real technical skills and it needed a lot of work from the climbing team to make sure that the Sherpas were safe. Hall and Fischer's commercial expeditions were very different. All the paying clients wanted, and expected, was to get to the top and the role of the Sherpas and the guides was to facilitate this. By now there were many Sherpas who were very strong and competent climbers in their own right and, as well as carrying loads and cooking, their role was to work alongside the mountain guides, shepherding the clients up what they nicknamed the 'yak route'.

There were obvious dangers here: if a climber wasn't capable of getting up a mountain on their own, should they really have been attempting it? And if not, then why were the companies so willing to take them to Everest?

EQUIPMENT

Most of the equipment used in 1996 was not fundamentally different from the gear that was taken by Hunt's team in 1953. Clothes and sleeping bags were warmer and lighter but they still worked on the same principle. Their tents, crampons and ice axes might have been made from more high-tech materials, but they were only more specialised developments of what had come before. The really big difference was in their communications equipment: in 1953 the British team had a radio set to listen to the World Service, but they couldn't make calls out and their triumphant news was taken by a runner from the foot of the mountain to the nearest radio-telegraph station. In 1996 by contrast, both teams had powerful walkie-talkies that allowed the leaders to keep in touch with their base camps. They had also both brought satellite telephones and one member of Fischer's team even had her own personal phone carried all the way to Camp 4 on the South Col. The age of the Internet had just dawned and through a cumbersome system of relays via fax and telephone messages, there were no less than five Internet 'broadcasts' going on from base camp.

CO-OPERATION BETWEEN TEAMS

In addition to Fischer's and Hall's, there were many other teams at Everest base camp that year, attempting to climb Everest from the Nepalese side. The roll call included two solo Scandinavian expeditions, Everest's first South African expedition, three further commercial outfits and a team hoping to haul a 42-pound (19-kilogram) camera to the summit to make an Imax movie. In order to avoid treading too much on each other's toes, several expeditions got together to arrange a tentative schedule for summit attempts. It was agreed that Fischer's and Hall's team would both go on the same day, 10 May. Two years earlier all seven members of another of Rob Hall's expeditions had reached the summit on 10 May and Scott Fischer had made his first ascent on 9 May 1994.

This may seem like a large number of expeditions but by the mid-1990s it had become the norm on Everest. In 1995, 150 people had reached the summit. Somewhere, there was an accident waiting to happen.

THE EVENTS UNFOLD

After about a month at base camp, during which time everyone got used to their equipment and made acclimatisation climbs, both teams prepared to leave for the South Col on 6 May. The assumption was that for the next stage, the final climb to the summit, all the clients would be using oxygen. Rob Hall and Anatoli Boukreev, a very strong Russian mountaineer who was employed as a mountain guide on Scott Fischer's team, had both climbed Everest without oxygen. However, none of the clients was strong enough and the prevailing orthodoxy was that, for the sake of maximum alertness, all the guides should also use oxygen on the summit day.

On 9 May both Hall's and Fischer's teams were camped high on the South Col, hoping to make an attempt the following day if the weather held. They were joined by the Taiwanese team, now reduced to a single climber, and the South African team. In total there were over 50 people camped on the Col, the windblown strip of purgatory that divided the Lhotse and Kanshung faces.

Ahead of them lay the South East Ridge, a mile-long ramp which leads up to the summit of Everest. Some of it was easy ground, but there were steep ice slopes and awkward patches of rock to get past. After passing the South Summit, a small rocky outcrop 300 feet (100 metres) from the top, the climbers would be faced with the final challenge, the Hillary Step, a 40-foot (12-metre) cliff named after guess who. After that, there was just a short slog to the small platform at the top. If Everest had been 10,000 feet

(3,000 metres) lower all of this would have been relatively easy climbing, but 29,000 feet (8,700 metres) up in the clouds, the battle against altitude made getting up the Hillary Step a real struggle.

In the early hours of 10 May, 33 climbers prepared to leave the South Col for the summit. Rob Hall's team was the first out: eight clients, three guides and four Sherpas, in theory a very safe ratio. Half an hour later, Fischer's team followed them. There were only six clients now, the two others having decided not to make an attempt for medical reasons. In support there were two guides and five Sherpas, again quite a comfortable ratio. Hall and Fischer had previously agreed to send some of their Sherpas ahead to lay fixed ropes along the most difficult parts of the ridge, but this didn't happen straight away. By the middle of the morning, the absence of these ropes was causing queues at the most difficult sections of the climb. Eventually the ropes were put in place, but what initially seemed like a minor misunderstanding ended up causing big delays on the mountain. Like many accidents, the disaster to come was not caused by one single catastrophic event; rather it was the build-up of a series of small mistakes like this, which snowballed into a much greater crisis.

Rob Hall had lectured his clients on the need to have a 'turn-around point', a time at which they should go back down if they hadn't reached the summit. Anyone who missed their turn-around point risked descending in the cold and dark. Nominally, it was supposed to be between 1.00 p.m. and 2.00 p. m. in the afternoon, but Hall never made this absolutely clear. At 11.30 a.m. three of Hall's clients did in fact turn around, realising that, because of the slow start and all the congestion, they would not be able to make it to the top and get back safely. Another client had turned back a few hours earlier. Scott Fischer's six clients were going slightly better, all of them doggedly heading on up.

Jon Krakauer reached the summit just after 1.00 p.m. Looking around him he saw the first wispy clouds, which presaged the coming storm. On his way down he had to wait for over an hour at the Hillary Step to allow a queue of climbers to come up. By 2.30 p.m. most of Scott Fischer's clients had reached the summit, but it didn't take long for the stress and strain to show on some of them. One had an injection of Dexamethasone, a powerful steroid, hoping that it would help to keep her going, another took a wrong turning at the South Col and then started to head down the precipitous Kanshung Face into China, before he realised his mistake.

In spite of all the lectures that he had given on the importance of turn around time, at 4.00 p.m. Rob Hall was still on the summit of Everest, waiting for his last client to struggle up the ridge. It was Doug Hansen, a Seattle postal worker, who had made his first attempt with Hall the previous year. In 1995 Hall had ordered him to turn back at South Summit because of bad snow conditions above, but he had encouraged Hansen to come back for a second attempt in 1996. Now Hall was waiting on the summit, knowing that they were both long past a safe turn-around time. Why did he ignore his own safety advice? Did he feel that he owed it to Hansen to help him reach the summit?

Hansen eventually did get there, but as he began to descend, his oxygen suddenly ran out and he found it almost impossible to carry on. Hall radioed down to ask for some fresh canisters to be sent up but one of his guides, Andy Harris, radioed back to say that the bottles which had been left at a depot on the nearby South Summit were all empty. In fact this wasn't true, and it is unclear why Harris made this mistake. Krakauer later speculated that Harris' own oxygen set was faulty, and that he might have been suffering from altitude sickness. Hall helped Hansen down to the Hillary Step but at this point he stopped, knowing that it would be impossible to go any further without fresh oxygen.

Scott Fischer, the other team leader, was also struggling. Even though he'd felt sick on the previous day he decided to make an attempt on the summit on 10 May. He made it to the top, but on the way back he started to pay the price. He managed to get down the Hillary Step, but after that he started to find things increasingly difficult. His friend and chief Sherpa, Lopsang Jangbu, tried to take him down, but Fischer was too heavy, so Lopsang left him on a protected ledge and then descended to get help. Makalu Gau, the Taiwanese climber who'd reached the summit that day, was also fading fast. His Sherpas roped him to the stricken Scott Fischer and then went down for reinforcements.

What had started as an orderly climb to the summit had, by the early evening, descended into chaos. A full-blown storm had erupted and visibility was appalling. With Fischer and Hall trapped on the mountain, both expeditions were effectively leaderless. But that wasn't the only problem. Hall had left two Sherpas in reserve at the South Col to help out if there were problems, but they had been knocked out by a build-up of carbon monoxide in their tents, caused by a problem with their stove. Anatoli Boukreev, the Russian guide who was working for Fischer's team, had earlier that day reached the summit without oxygen and then climbed quickly down to the South Col to warm up. He was prepared to climb back up with fresh canisters for anyone who needed them, but he didn't have a radio so he didn't know what was going on.

By 9.00 p.m., 19 of the 33 climbers who had set out were missing. A large group made it down the South Col and then got lost in a blizzard. Unable to find the tents, they were forced to huddle down in the snow to wait for the weather to break. After several hours, some of them staggered into the camp; Anatoli Boukreev heroically ventured out to rescue the others but by morning there were still several people unaccounted for.

Rob Hall survived a night out on the mountain, and the following day he was able to contact his base camp by radio. He told colleagues that his client Doug Hansen was dead and that the other guide Andy Harris was missing. He said that he was having problems with his oxygen set, but that soon he was going to make an attempt to come down. That day two calls were patched through to him via satellite phone, from his wife, thousands of miles away in New Zealand. He repeated that he was about to move off but he never managed to. Two Sherpas went up in a desperate attempt to rescue him, but the winds were too strong and they didn't even make it as far as the South Summit. Another pair of Sherpas tried to rescue Scott Fischer and the Taiwanese climber, but Fischer was in such a bad way that they were forced to leave him there. When Anatoli Boukreev reached him several hours later, he was dead.

THE DEAD

Ultimately four members of Adventure Consultant's expedition died: two clients, Doug Hansen and Yasuko Namba and two guides, Andy Harris and Rob Hall. All of Scott Fischer's team survived except for the leader himself. There had been far worse disasters before, but the events of 1996 received unprecedented publicity and were raked over endlessly in the press, on television and on the Internet. The question that everyone wanted answered was simple: where had it all gone wrong?

A lot of articles in the press focused on the commercialism of the expeditions: was the main qualification now for climbing Everest a healthy bank balance? In the past, veterans such as Edmund Hillary and leading climbers such as Reinhold Messner had criticised commercial expeditions for devaluing Everest and for allowing clients to get on to mountains that they weren't really capable of handling. Two of Hall's clients died and it was a miracle that all of Scott Fischer's people had survived. Had commercial pressures clouded the judgement of the two leaders? Had they allowed clients who weren't really up to it to go on their expeditions?

When the stories came back of crowds of climbers queuing up at the Hillary Step and millionaire socialites having copies of *Vogue* sent to them at base camp, it all seemed to be a far cry from the heroic days of Mallory and Irving. The overcrowding wasn't just a sign of decadence though; it caused real problems on the mountain, slowing everyone down and therefore increasing the risks. Afterwards, Jon Krakauer fancifully suggested banning oxygen from Everest altogether, in order to raise the stakes and keep weak climbers away. It is unlikely that this suggestion will ever be taken up.

There were clearly bad judgement calls from both of the expedition leaders. Rob Hall lost his life by refusing to leave a sick client on the mountain, but he only got into that situation by breaking his own rules. It would have been a hard call, but if he had made Doug Hansen stick to a realistic turn-around time, then both men might still be alive. Did he feel guilty about the idea of asking Hansen to turn back, for the second time in two years, knowing how much Hansen had invested in getting to the summit? Scott Fischer paid Anatoli Boukreev $25,000 to work as a guide on his expedition, yet he allowed the Russian to climb without oxygen on the summit day and then go down ahead of everyone else to the South Col. Wouldn't it have been better to insist that the Russian use oxygen and stay with the clients? Fischer told several people that he was feeling ill on the summit day, but he still carried on to the top. If he had stayed in his tent that day, he too might have survived. Did he feel a commercial pressure to make the summit in order to get more clients in the following year?

It is well known that accidents happen on the descent: climbers are tired and flushed with success and they often let their guard down. At high altitude all these problems are compounded by the steady ticking of the altitude clock. Rob Hall's clients were each given three oxygen cylinders on summit day; this gave them roughly 18–20 hours' worth of air. As the delays mounted on the way up because of all the overcrowding, the climbers used up their oxygen just waiting around. It is not surprising that on the way down, many of them ran out of gas.

None of the deaths in 1996 was caused by simple human error. The climbers were killed by the cold, the altitude and the capricious weather. Mountaineering is inherently dangerous and high-altitude mountaineering even more so. Within a few years Anatoli Boukreev was killed on another Himalayan mountain, Annapurna. Climbing is a risk sport and the higher you go the higher the risks become but . . . that doesn't mean that those risks cannot be managed, and mismanaged. Climbing to the top of the world is always going to be difficult; getting to the top of the world with 32 other people on the rope behind you is a lot harder.

LUCK

Victory awaits him who has everything in order – luck people call it. Defeat is certain for him who has neglected to take the necessary precautions in time – this is called bad luck.

ROALD AMUNDSEN

If you look at expeditions that have ended in disaster, it is often easy to see the malign hand of fate taking a prominent role. There are, invariably, so many 'what ifs': if Burke had arrived at Cooper's Creek just a few hours earlier, if Apsley Cherry Garrard had taken the huskies just a few miles further on, if Doug Hansen had turned around before he reached the summit . . . it could all have been so different. Exploration by its very nature is a leap into the unknown; explorers and mountaineers are permanently at the mercy of the weather which can make or break an expedition. *But* . . . as Amundsen wrote in his account of his South Pole expedition, to a certain extent everyone makes their own luck. If Burke had turned around earlier, he would have reached Cooper's Creek earlier; if Scott had given himself a greater margin for error, he wouldn't have suffered from such a shortage of fuel; if Rob Hall had enforced a strict turn-around time he would have got down earlier . . . unless some unforeseen disaster occurred.

The two very simple things that will bring success much closer are good preparation and good judgement. Both of these skills are partly instinctive and partly learned through experience. In the end, luck may play a large role in all expeditions, but planning, leadership and good teamwork make bad luck much easier to cope with, and good luck much easier to profit from.

ode

Rev

Safety

Rep

Triumph

Examine Bio

ame Glory

Press Fake

Supporter Scand

Controversy D

Exhaust Warri

Proof

Fre

Getting Back

The World is a book, and those who do not travel read only a page.

ST AUGUSTINE

TELLING, SELLING AND TRYING TO
RETAIN CONTROL OF YOUR OWN STORY

Towards the end of many expedition diaries, thoughts turn to home and the fastest way to get there. There are hands to shake and partners to hug and meals to eat. Some explorers and mountaineers find it difficult to adjust to 'normal' life, but most seem to relish a little bit of domesticity, for a few weeks at least. Then, almost invariably, they have to get out their quills, typewriters and word processors to starting writing their story. It is remarkable how quickly expedition books get into print: John Hunt's account of the 1953 Everest expedition was in the shops barely four months after he returned from an utterly exhausting expedition. Obviously he wrote it quickly, because there was such huge interest in the expedition, but every explorer knows that the best way to keep control of your own story is to get into print straight away. If you don't then there is always the risk that someone else will get there first.

SHAMELESS DIARIES AND
HUNGRY PRESSMEN

In 1903 Robert Dunn, a young journalist, joined the American explorer Frederick Cook on his first expedition to Alaska. Cook was by then quite a well-known figure, having made his name on a number of polar expeditions; this time he was aiming to climb Mt McKinley, the highest peak in North America. Dunn went along on the recommendation of Lincoln Steffens, the editor of one of Randolph Hearst's newspapers. Steffens was instinctively suspicious of all the attention that explorers were getting and was convinced that if an independent eyewitness was present, the dark truth of expedition life would be revealed with all its arguments and petty bickering.

Robert Dunn's sensational expedition diary was eventually published in an outdoors magazine and it later became the basis of his splendidly titled book, *The Shameless Diary of an Explorer*. In fact the book itself wasn't quite as cynical as its title suggested. By the end of the expedition Dunn had warmed to Cook and even found something to admire in him, even though he had failed in his stated intent to climb Mt McKinley. The story behind *The Shameless Diary*, however, illustrates the ambiguity of the press towards explorers: they are as happy to damn them as they are to praise them. This wouldn't be very important if it wasn't for the fact that over the last two centuries the press and the other media have become so important to exploration.

From the mid-nineteenth century onwards, exploration started to become as much a symbolic activity as one that had a utilitarian value. In Britain, mountaineering became popular in the Victorian era, around the same time as the rules were being formalised for other 'sports' such as rugby and cricket; some climbers liked to claim a scientific or mystical dimension to their chosen sport, but it was clear to most people that mountaineering served no practical purpose other than to get jaded city dwellers out into the fresh air and give them plenty of satisfaction. Polar exploration was usually carried out on a grand scale, but it too was seen by many as 'pointless': as early as the eighteenth century you can find articles and cartoons in the British press, lampooning various explorers' obsession with the North Pole. In the sixteenth century there was a commercial purpose for trying to find a Northwest Passage, but by the nineteenth century it too was seen as very much a symbolic prize. To many explorers, none of this dampened their passion, but the problem with activities which had no practical value was that it was difficult to raise money for them and, as every explorer knew, filling in blanks on the map was an expensive passion.

And so the supporters of exploration started to bang a different drum; instead of claiming that there was any good practical reason for supporting another expedition to the North or South Pole, they started to claim that it was good for national prestige and that there was something inherently noble in taking on one of Nature's challenges. Would the public buy into this vision though? The answer was yes.

They were clearly fascinated; they wanted heroes and, sitting in the comfort of their armchairs, they loved to hear all the exotic stories and gory details from expedition life. Mountaineers and explorers rushed into print. It was a new way of funding their adventures and gaining public recognition. Even in the 1860s, the American explorer Elisha Kent Kane was able to pre-sell 20,000 copies of his account of a recent Arctic expedition. In 1929 the Australian explorer Sir Hubert Wilkins wrote a book about his proposed journey under the Arctic ice in a submarine, *before* the expedition actually took place, because he didn't have the money to fund it. The expedition book has increasingly become part and parcel of the explorer's life; the British climber Eric Shipton wrote seven books and Reinhold Messner has written no less than 17.

TEN FAMOUS EXPLORATION BOOKS
* * *

Scrambles Amongst the Alps

Edward Whymper's story of his early climbs in Switzerland and the accident on the Matterhorn. There were six British editions in Whymper's lifetime and 11 further foreign editions.

Arabian Sands

Wilfred Thesiger's account of five years in the Arabian Peninsula is widely thought of as one of the greatest books ever written on desert exploration.

Touching the Void

Joe Simpson's story of how he struggled to stay alive after a climbing accident in Peru. It sold a million copies before a film version was released and has sold many more since.

The Worst Journey in the World

Apsley Cherry Garrard's account of Scott's second expedition and his own journey to find Emperor King Penguin eggs. It deserves its fame for its title alone and is still in print.

How I Found Livingstone

Henry Morton Stanley's detailed account of his search for David Livingstone was a success in its day in spite of its 700 pages.

Annapurna

Maurice Herzog's account of his dramatic Himalayan expedition in 1950. It is said to be the most successful mountaineering book ever. Herzog didn't benefit directly from the book: the profits went to the French Alpine Club.

Into Thin Air
Jon Krakauer's dramatic re-telling of the 1996 Everest Disaster. It is one of the most successful adventure books of recent years.

Seven Years in Tibet
Heinrich Harrer's account of his life in Tibet and his friendship with the Dalai Lama. It is reputed to have sold ten million copies.

The Valley of the Assassins
Freya Stark's vivid description of her travels in Persia won her immediate acclaim and revealed her as a very idiosyncratic voice.

Scott's Last Expedition, The Journal of Captain R. F. Scott
Captain Scott's journal was published just a few months after his body was discovered in the Antarctic. Certain comments on the members of his team were expurgated.

There is no doubt that literary ability plays a big part in the long-term success of an expedition book. In the short term though, if your story has made the news, then copies will shift regardless of how well or badly it is written. Scott's journals were rushed into print, and sold by the thousand; in fact they are very well written, and Scott's eloquence certainly enhanced his reputation. His rival, Roald Amundsen, was not a particularly good writer and his account of the successful Norwegian South Pole expedition was very dry and unexciting. Whatever your literary ambitions, there are certain things you can do to improve your chances of publishing success.

HOW TO WRITE AN EXPLORATION CLASSIC
Do something amazing and then write about it.

Suffering sells There is nothing like a good frost-bite story or near-death experience to shift an exploration book. Painography is the main selling point of many mountaineering books.

Come up with a good title Eric Shipton's *Blank on the Map*, Apsley Cherry Garrard's *The Worst Journey in the World*, Joe Simpson's *Touching the Void*: these were all great titles which really helped to sell the books. However, Knud Rasmussen's

Report of the Fifth Thule Expedition, Volume 8 is a classic of put-downable titles.

Don't use a ghostwriter There are very few expedition classics written in cahoots with a ghostwriter. Robert Peary used one for his later books but they were uniformly dull. The American novelist James Ramsay Ullman ghostwrote Tenzing Norgay's first 'autobiography', *Man of Everest*; it sold well initially but many reviewers were suspicious of Ullman's role.

Keep it simple A bare, stripped-down style seems to suit exploration books. Great explorers such as Wilfred Thesiger and Eric Shipton were masters of Spartan prose.

Keep it short There are five volumes in Heinrich Barth's account of his travels in Africa, two volumes in Amundsen's account of his expedition to the South Pole and nine appendices in John Hunt's slim account of his Everest expedition. Explorers seem to feel a deep need to mirror their enormous achievements with enormous amounts of words but how many readers reach the final page?

Avoid writing dialogue There is nothing more tedious than badly written dialogue, especially when it takes place at some remote base camp in the middle of nowhere.

Dealing with the Press

The media play a key role in the fame game so it is important to know how to deal with them. Good coverage can bring welcome publicity at the outset of an expedition and a nice pat on the back at the end, but it is always worth remembering that the media often has its own agenda which colours the way a story is reported. The 1953 Everest expedition is a good example of an expedition which got most things right, except its relations with the press.

Everest beats the climbers
The Evening Standard, 28 May 1953

British fail in attempt to climb Mt Everest
The New York Herald Tribune, 29 May 1953

Everest men return, base camp again
The Daily Telegraph, 30 May 1953

The top of Everest
The Times, 3 June 1953

Tenzing will not come to Britain
The Daily Mail, 22 June 1953

'First up' row over Everest
The Daily Express, 22 June 1953

Tenzing is reported to be miffed
The New York Herald Tribune, 23 June 1953

Tenzing under armed guard
The Daily Express, 23 June 1953

Tenzing appeals: End this bickering
The Daily Mail, 29 June 1953

The die was cast in January 1953 when the organising committee signed a deal with *The Times*, trading exclusive access to the story of the expedition for £10,000. There was a long history of involvement with *The Times*: in the 1920s and 1930s, it had sponsored all but one of the British expeditions to Everest. It seemed like a very good deal but the organisers had underestimated the worldwide interest in the expedition.

When the British team arrived in Kathmandu in April 1953, Indian journalists were keen to talk to John Hunt and his men, but Hunt stuck rigidly to *The Times* contract and, bar one press conference, he refused to authorise any interviewees. This was a huge frustration: Indian and Nepalese journalists regarded Everest as *their* mountain, and the Sherpas were *their* men, so why couldn't they cover the expedition properly? India had only recently gained its independence from Britain and its journalists were not so keen to kowtow to their former masters. Tenzing was the only member of the expedition who was willing to give interviews; he refused to sign the same non-disclosure contract as the other British climbers. So, partly because they had access to him, the Indian press began to build Tenzing up as the real force behind the expedition; he had got very high on Everest with a Swiss expedition in 1952, and now he was going back to the mountain to finish the job.

James Morris, *The Times*' special correspondent, spent a lot of his time trying to work out how to get his exclusive reports back home quickly. He was the only journalist allowed to travel with the expedition but it was not going to be easy to stay in touch with Fleet Street. Today satellite phones and Internet links are available to well-booted expeditions, but the 1953 British team didn't even have a radio transmitter. Morris's reports were literally carried back to Kathmandu by waiting runners; this meant a news lag of four to seven days and the inevitable risk of his scoops being stolen.

The Times' rivals were never going to respect its exclusive contract with the expedition. On the way out, James Morris bumped into Ralph Izzard of the *Daily Mail*, returning from an unauthorised visit to base camp. His solitary trek had taken him 20 days there and he hardly got anything out of the British team, apart from a cup of tea and a general discussion of the weather. Nevertheless his 'epic' trip out, shod in plimsolls, generated several articles for the *Mail* and a book, *The Innocent on Everest*. Today the idea that someone should make a solo trek all the way out to base camp without getting even a brief interview seems rather mean-spirited and slightly absurd, but everyone in the expedition had signed a contract and they were sticking religiously to it.

By the beginning of May, Kathmandu had filled up with a large international press corps, hungry for a story. A few journalists ventured out of the city, but most preferred to spend their time attempting to bribe Morris's runners and endlessly speculating on the fate of the expedition. At the end of the month, British papers were full of alarming reports that the expedition had failed. Their authors invariably cited sources 'close to the expedition' and mysterious radio voices. In fact it was all just guesswork: John Hunt, the expedition leader, had said that he wanted to be ready to make a first attempt by the middle of May so with no news in Morris's postbag, everyone had assumed the worst. There were so many of these reports that, back in London, the expedition's organisers started to get jittery. They drafted a press release to announce that the expedition had failed and booked the tickets for a second team to go out in the autumn.

In the end they never put out the press release or picked up the tickets. On the night of 1 June, the eve of the coronation, a message reached London from James Morris with the great news that Hillary and Tenzing had reached the summit. Two days earlier, by a stroke of luck, James Morris had been at advance base camp to witness the arrival of Hillary and Tenzing. He rushed down the mountain, hammered out a coded despatch and handed it to a runner on the following day. Morris had discovered that there was an Indian army radio station 30 miles away from the foot of the mountain and had persuaded the operator to send messages for him. In order to minimise the risk of leaks he decided to use the station just once, for the announcement of failure or success.

At the beginning of the expedition, Morris had devised an elaborate code to confuse any spies, attaching a phrase to each climber's name. The beauty of it was that the encoded message didn't look unusual. The final message, announcing the success of the expedition, read: 'Snow conditions bad (*Everest climbed*), advance base abandoned (*Hillary*), awaiting improvement (*Tenzing*). All well.' When it reached the embassy in Kathmandu it was deciphered by Christopher Summerhayes, the British ambassador, and forwarded to the Foreign Office in London. They passed it on to *The Times*, who then published the news in a special coronation edition that evening. On the following morning the Everest story was spread over the front pages of most of the British newspapers. They credited American television as their source, flagrantly ignoring the fact that it was James Morris's hard work that had got the news out so quickly.

The story might have ended there but the polarised reporting that characterised the beginning of the expedition now returned at its climax. When John Hunt reached Kathmandu in the middle of June, he found himself in the middle of an unexpected controversy. For the last fortnight Indian newspapers had been reporting that Tenzing had beaten Hillary to the summit. No one had actually interviewed Tenzing, but as usual 'sources close to the expedition' had confirmed the story.

When Hunt returned to Kathmandu with an advance party, he was asked to comment; he denied that this was true and insisted that the whole issue of who stepped on the summit first was irrelevant: Hillary and Tenzing had been roped together and they were climbing as a team. When Hunt was then asked how Tenzing rated in comparison to a Swiss mountain guide, he was uncharacteristically undiplomatic. He said that though Tenzing was a talented mountaineer he just wasn't in the same league. Technically speaking this was true but, as Tenzing later remarked, comments like this only poured petrol on to an already raging fire.

When Tenzing himself finally arrived in Kathmandu a week later, he was treated like a returning god. On the outskirts of the city he was asked to sign documents confirming that he had reached the summit first; Tenzing couldn't read and he later said that he wasn't aware of what he was signing. The streets were adorned with posters, showing him triumphant on the summit whilst Hillary clung on somewhere below. Both Indian and Nepalese journalists tried to claim Tenzing as their own national hero but they were united on one thing: he had reached the summit first.

For the British journalists who were still in Kathmandu, this controversy was seen as a great opportunity to keep the Everest story running and make mischief. *The Times* still had the exclusive rights to cover the expedition, but it didn't have a monopoly on the ensuing controversy. The *Telegraph* attacked the British ambassador for using diplomatic channels to transmit a message for *The Times*, the *Mail* serialised Tenzing's inside story of the expedition and the left-wing *Daily Worker* criticised Hunt for his disrespectful treatment of Tenzing. *The Times'* man in Kathmandu, Arthur Hutchinson, vainly appealed to his bosses back in London to allow greater access to the team in order to clear up the controversies but they refused. They had syndicated the Everest story all around the world on the basis that they had exclusive access to the team and they weren't going to give it up now.

The climbers themselves reacted to all the fuss with weary bemusement. It had taken a huge effort from all the team and all the Sherpas to put Hillary and Tenzing on to the summit, but as far as most of the press were concerned none of them mattered. Edmund Hillary was aware of the controversy over who reached the summit first and it rankled. However, when he came to write an article for *The Times* on the final stage of the expedition, John Hunt and the ambassador Christopher Summerhayes persuaded him to avoid the issue. In his original version, he described how he had reached the summit and then brought Tenzing up, but in the published version he made a more neutral reference to the moment when 'we' reached the summit.

For Tenzing it was all too much. He was constantly invited to awards ceremonies and official celebrations, but he rarely looked happy. Somehow he had to steer a course between competing loyalties to the team, to India, his adoptive country, and to Nepal, the country of his birth. He too leapt on the formula 'we reached the summit together' which was eventually accepted by most of the press. As soon as that controversy died down, others came to take its place. Many papers speculated on whether Tenzing would accompany the team back to London; Hillary and Hunt had been awarded a knighthood but Tenzing had to make do with the George Medal, apparently because the Indian government didn't want him to be knighted. Tenzing was reported to be angry about John Hunt disparaging his mountaineering skills; after all, he said, he had just climbed the highest mountain in the world.

Today in India you can still find people who insist that Tenzing reached the top first even though he himself later confirmed in his autobiography that Hillary was in the lead. It is obvious that the whole controversy was cooked up by the press. Long before they spoke to either man, Indian newspapers were reporting that Tenzing had reached the summit before Hillary. This kind of partisan reporting happened partly because of postcolonial assertiveness and partly because Indian journalists felt that when the news was first reported on coronation day, British newspapers had paid too much attention to Hillary. It didn't help though that the expedition handled the press so badly. If they had let the Indian and Nepalese press in at an early stage, their first reports would not have been so polarised. If they had talked to them more at the end, then it's unlikely that the controversy would have lasted so long or been so bitter. If they had given the British Press regular briefings then they would not have been so keen to publish rumours and guesswork. If the organising committee had understood the mood of the country, and indeed the mood of the whole world, they would have realised that they needed to be much more open-minded.

Today all the arguments at the end of the expedition are largely forgotten; in the grand scheme of things they are a very small part in the story of a very successful expedition. At the time though, all the bickering was the last thing that Hunt, Tenzing and all the other climbers needed at the end of an exhausting expedition. The lesson from this episode is that you can't dictate to all the press how your story will be written, so you need to be sensitive and sophisticated in the way you deal with them. Hell hath no fury like a journalist scorned.

PROVE IT

At the strangest and most hysterical point of the Everest controversy, an Indian journalist published a series of articles claiming that, in fact, the British had not reached the summit and the whole story was a fraud. He maintained that James Morris's final message hadn't been in code at all and that the expedition film had in fact been shot in a studio in Calcutta! This seems like an archetypal twentieth-century conspiracy theory, but in fact there is a long tradition of exploration controversies which have frequently broken out in public. In the 1820s Captain James Clark Ross was openly attacked in print, by the men who had served under him on an

Arctic expedition, over the accuracy of his maps. And to cap it all off, one of his staunchest critics was his own nephew.

Perhaps the most vicious dispute of all took place in 1909 when the American explorers Robert Peary and Frederick Cook slugged it out in the press over their rival claims to the North Pole. Both men were well-known and highly regarded figures. Peary had spent almost two decades in the Arctic and had already made several attempts on what he called the DOP, the 'Damned Old Pole'. His final expedition was one of the best-prepared and best-organised polar expeditions ever. When he returned from the Arctic in September 1909 to announce that he had finally reached his goal no one was surprised. The only problem was that another American, Frederick Cook, was claiming to have beaten him to it.

Cook was a remarkably charismatic man; he was a doctor by trade, but an explorer by inclination. He first went to the Arctic as the expedition doctor on one of Peary's first forays into Greenland. After falling out with Peary over the right to publish an article about the expedition, Cook went his own way. He led two expeditions to Alaska, returning in triumph in 1906, after his second expedition, claiming to have made the first ascent of Mt McKinley, the highest mountain in North America. When in September 1909 Cook announced that he had just come back from the North Pole, initially no one doubted him – until Peary got back to America.

For the next few months the two men and their respective supporters battled it out in the newspapers. The problem for both explorers was that there was no real way to prove that either of them had been to the Pole. They each had photographs of the moment of their triumph, but all that could be seen was a vast expanse of white ice behind them. There is no land at the North Pole: it's a geographical abstraction located in the middle of the Arctic Ocean. When the weather gets cold enough the sea freezes into hard plates of ice which float over the Pole; at other times of the year it is simply water. Both men had notebooks full of navigational readings which they had used to plot their position in the Arctic but as both sides pointed out, these could have been faked. In the end it came down to whose word you believed; to many, Cook seemed like the underdog and Peary the classic bully with lots of friends in high places.

Peary played it tough. He brought back an affidavit from the Eskimos who were supposed to have accompanied Cook, denying that they had gone anywhere near the Pole. His supporters sent private investigators on Cook's trail. Then suddenly Cook's story started to implode. Peary's supporters tracked down the man who had accompanied Cook on his ascent of Mt McKinley in 1906, and like the Eskimos in Etah he signed an affidavit affirming that he and Cook had not reached the top of the continent. Again, some of Cook's supporters put this down to bullying and bribery but the man, Ed Barrill, also provided a map showing where they actually went in Alaska and added that the famous summit photograph was a fake. A group of mountaineers headed north with Barrill's map and came back with a set of photographs which were almost identical to those taken by Cook, but which weren't taken anywhere near McKinley. Cook had been hoisted with his own petard: if he hadn't come back with any photographic evidence of his climb, he could have toughed it out and argued convincingly that, as Barrill had been paid thousands of dollars by the Peary camp, he couldn't be trusted as an impartial witness.

Within a few months of the McKinley exposé, Cook had disappeared and no one apart from a die-hard rump of supporters believed him. Ironically though, Peary's pursuit of Cook now put the spotlight on his own claims and in certain respects he too was found wanting. The only witnesses to his conquest of the North Pole were two Eskimos and his assistant Matthew Henson, but none of these had taken any navigational readings. It turned out that Peary was claiming to have travelled astonishingly quickly, that he had hardly taken any readings for longitude and that he had spent much of the journey navigating by 'dead reckoning'. Peary was very cagey about his navigational data and refused to hand over his diaries to an official committee. In his favour, unlike Cook, Peary had had a very strong support team and a feasible plan, and he didn't have as many skeletons in the closet as his former colleague, but the lingering suspicion that he hadn't gone all the way to the Pole never went away. In 1984, the British explorer Wally Herbert was given exclusive access to Peary's diary for a piece that he was writing for *National Geographic* magazine, a long-time Peary supporter. Instead of ending the controversy, Herbert's article only deepened it. He discovered that the three pages covering the time that Peary spent at the Pole were blank. This seemed to add to the evidence that Peary had not told the truth, but Peary's supporters rejected this idea, and another article soon appeared in *National Geographic*, once again proving that Peary had got there.

The dispute over the North Pole and the revelation that Cook had faked his ascent of Mt McKinley gave the public more reason to be sceptical about the 'heroic' exploits of gentlemen explorers. Paradoxically, the arrival of photography, which had seemed to offer the possibility of coming back with unequivocal proof, simply added a new level of complexity to the whole issue of proof. When Cook's detractors got hold of his original photograph of the summit of McKinley, they discovered that it had been cropped and the original negative showed a couple of distant peaks in the background which should never have been there if this truly was the summit. Using the map supplied by Ed Barrill, they were able to find the exact spot where Cook had taken the picture; it was duly noted down on their map as 'Fake Peak'. The moral of the story for any future fraudsters is to keep your proofs as vague as possible.

In recent years, there has been a lot of interest in finding a camera carried by George Mallory on his attempt on Everest in 1924. Mallory and his partner Sandy Irvine famously disappeared on their summit attempt, leaving open the question of whether they had gone all the way to the top. Several expeditions have gone out to look for Mallory's camera, hoping that if it was recovered, the negatives inside might be able to solve the mystery. Apart from the low odds that the camera will ever be found, it's wishful thinking to assume that any photographs will resolve the arguments for once and for all.

What Next?

Like James Dean and Buddy Holly, George Leigh Mallory sealed his reputation by dying young and leaving more questions than answers. Knowing when to stop and finding something to do next have always been difficult issues for explorers and mountaineers. There is no obvious career to move into, after spending long years of your life living out of a rucksack. Several explorers entered politics, or tried to: Reinhold Messner became an MEP for the Green party in the European Parliament; Maurice Herzog became a minister in De Gaulle's government and was for many years the Mayor of Chamonix. Henry Morton Stanley tried but failed to be elected as a Liberal MP and Ernest Shackleton failed to be elected as a Unionist MP. Fridtjof Nansen was invited to work for the League of Nations after the First World War and Edmund Hillary spent four years as the New Zealand High Commissioner to India. After his success on Everest, Sir John Hunt spent many years in public life, chairing the Duke of Edinburgh Award Scheme and the Parole Board.

A lot of explorers, though, never found a place for themselves in civvy street and just kept on going until they dropped.

DIED IN THE FIELD

David Livingstone died in Africa in 1873, two years after he was famously rescued by H. M. Stanley. His heart was buried in Africa, the rest of his body embalmed before being sent back to Britain.

Tom Bourdillon almost made it to the top of Everest in 1953. He died in 1956 in a climbing accident in the Alps.

Gino Watkins, the leader of the British Arctic Air Route expedition, died in 1932 whilst hunting seals in Greenland.

KILLED BY NATIVES

Captain Cook, the British explorer, was killed by a Hawaiian warrior on the west coast of the island. His body was hacked to pieces by a frenzied mob before being ritually cooked in honour of his great power.

Ferdinand Magellan, the Portuguese explorer, was killed by natives in the Philippines. Magellan didn't die easily: it took a poison arrow and several spears to finish him off.

Mungo Park, the Scottish explorer, was killed on an expedition to the Niger in West Africa. Having fended off attacks from herds of hippopotami, Park and his men were ambushed at Bussa Rapids on the Niger; his body was never found.

MISSING IN ACTION

Percy Fawcett disappeared in the jungles of Brazil in 1925. Over the next 20, there were reported encounters with strange characters who could have been Fawcett in the jungle.

Roald Amundsen disappeared off the coast of Norway in 1929. He had come north to join the international rescue expedition searching for the Italian, Umberto Nobile, whose airship had crashed in the Arctic. Ultimately Nobile was rescued but Amundsen's plane crashed and no bodies were ever recovered.

Bill Tilman, the legendary British yachtsman and climber who accompanied Eric Shipton on many of his expeditions, disappeared in the South Atlantic in 1977, when at the age of 80 he went on an expedition to the remote Smith Island in the Antarctic.

Of course, you don't have to go out with a big bang; many explorers and mountaineers had much quieter deaths than these and some really did just keep on going. Freya Stark, for example, carried on travelling well into her eighties. She died, aged 100.

*   *   *

So, the million dollar question – Why?

---

*If you need to know, you'll never understand and if you understand,
you'll never need to know.*

ANON

There is nothing more likely to irritate an explorer or a mountaineer than to ask them why they do it, or why they are so willing to put up with, what seems to many others, such danger and discomfort. The problem though, is that this is a question that just won't go away. So explorers have to become great stone-wallers: George Mallory's 'Because it is there' is one of the most famous quotations in the history of exploration. All his poor interviewer wanted to know was why he wanted to go back to Everest: couldn't he have appealed to the restless yearning of the human spirit, or man's eternal need to discover or the simple desire to be the first person to tread on the highest point of the earth? Perhaps, having been asked the same question several hundred times, Mallory just didn't care anymore, and 'Because it is there' was simply the first phrase to come into his head.

For armchair explorers and aficionados of adventure, 'Why?' has always been one of the most interesting questions. Some of the answers are fame and fortune; these are the goals of many other activities and have obvious attractions. Add a dash of patriotism and a smattering of 'the eternal quest of mankind' and you need look no further. But what are you to make of Wilfred Thesiger crossing the Empty Quarter for the sake of a sip of water, or Apsley Cherry Garrard risking his life and limb for the sake of half a dozen Emperor King Penguin eggs? The answer is that exploration is a very personal activity and there are many different motivations. Wilfred Thesiger's cup of water is a self-consciously insignificant goal which nevertheless speaks of the simplicity of his approach to exploration. A glass of water is both ubiquitous and perfect: what *could* be nicer at the end of a long journey? Apsley Cherry Garrard's penguin eggs are of a different order: though to the common man they might not seem like an end worth dying for, to anyone interested in science they are talismanic objects. When several months after the conclusion of the expedition Cherry Garrard goes to the Natural History Museum to present his penguin eggs and discovers that the curators are indifferent, you can feel his anger in a very visceral way. The deaths of Scott and the others on the same expedition have added considerably to the value of the eggs which become a kind of validation of the whole expedition.

On the other hand, perhaps, 'Because it is there' was simply a way of saying, 'Why not?' What might seem bizarre to a non-mountaineer seems self-evident to someone like Mallory. You climb Everest, because it is climbable and that's what you like doing. 'Because it is there' is not stonewalling – it's just stating the obvious. One very simple way to look at mountaineers and explorers, or successful ones at least, is to see them as people who have realised what they are good at. When you read books by Livingstone or Shipton or Peary or Heyerdahl, you get an overwhelming sense of men who are at ease with their environment, however it alien it might seem to an outsider. They do what they do because they have an instinctive grasp of it.

This obviously makes questions about motivation, in one sense, a little less interesting. It always sounds better if you have some quasi-mystical reply which, even if it doesn't really explain things, at least sounds good. Nevertheless it is fundamentally true that aptitude is a powerful motivator in itself. If you discover that you have no trouble at all climbing steep rock-faces or that you have no problem at all dealing with the dry heat of the desert or that you pick up languages easily, then it is not surprising that you should seek an outlet for your talents.

If there is one lesson to take away from this book, apart from how to make your underwear last and how to escape from an anaconda, it is that success and happiness are dependent on finding out what you are good at and then doing it. So go get lost, and then find yourself.

ℬibliography

Primary Sources

Andrée, Dr Salomon *The Andrée Diaries* Lane (1931)

Albanov, Valerian *In the Land of White Death* Modern Library (2000)

Amundsen, Roald *The North West Passage: Being the Record of a Voyage of Exploration of the Ship 'Gjöa' 1903–1907* Constable (1908)

Amundsen, Roald *My Life as an Explorer* Heinemann (1927)

Amundsen, Roald *The South Pole* Murray (1912)

Baker, Samuel *Ismailïa: A Narrative of the Expedition to Central Africa for the Suppression of the Slave Trade* Macmillan (1874)

Barth, Heinrich *Travel and Discoveries in North and Central Africa* Longman, Brown, Green, Longmans and Roberts (1857–1858)

Bates, Robert and Houston, Charles *K2, The Savage Mountain* Collins (1955)

Bates, Robert and Houston, Charles *Five Miles High* Robert Hale (1939)

Bingham, Hiram *Lost City of the Incas* Duell, Sloan and Pearce (1948)

Bonatti ,Walter *On the Heights* Hart-Davis (1964)

Bonington, Chris *The Next Horizon* Gollancz (1976)

Boukreev, Anatoli *The Climb* St Martin's Press (1997)

Burton, Richard *The Lake Regions of Central Africa* Longmans (1860)

Byrd, Richard E. *Alone* Putnam (1938)

Calvert, James *Surface at the Pole* Hutchison (1961)

Chapman, F. Spencer *Northern Lights* Chatto and Windus (1933)

Cherry Garrard, Apsley *The Worst Journey in the World* Constable (1922)

Compagnoni, Achille *Men on K2* Veronelli Editore (1958)

Cook, Frederick *Through the First Antarctic Night* Heinemann (1900)

Cook, Frederick *To the Top of the Continent* Doubleday, Page and Co. (1908)

Cook, James *Captain Cook's Journal During His First Voyage Round the World* (1893)

Crowley, Aleister *The Confessions of Aleister Crowley* Cape (1969)

Danielsson, Bengt *The Happy Island* George, Allen and Unwin (1952)

Desio, Ardito *The Ascent of K2* Elek Books (1955)

Douglas, Ed *Tenzing, Hero of Everest* National Geographic (2003)

Dunn, Robert *The Shameless Diary of An Explorer* Outing Publishing Co. (1907)

Evans, Charles *Kanchenjunga, The Untrodden Peak* Hodder and Stoughton (1956)

Fawcett, Percy *Exploration Fawcett* Hutchison (1953)

Freuchen, Peter *Book of the Eskimos* The World Publishing Company (1957)

Galton, Francis *The Art of Travel* Murray (1872)

Gibbons, A. St H. *Hunting and Exploration in Central Africa* Methuen (1898)

Hartley, Catherine *To the Poles Without a Beard* Simon and Schuster (2002)

Hempleman Evans, David *Walking on Thin Ice* Orion Media (1999)

Henson, Matthew *A Black Explorer at the North Pole* Brompton Books Corporation (1989)

Herbert, Wally *Across the World* Longman (1969)

Herzog, Maurice *Annapurna* Cape (1952)

Hesselberg, Eric *Kon-Tiki and I* George, Allen and Unwin (1951)

Heyerdahl, Thor *Fatu-Hiva* George, Allen and Unwin (1974)

Heyerdahl, Thor *The Kon-Tiki Expedition* George, Allen and Unwin (1950)

Heyerdahl, Thor *In the Footsteps of Adam* Little, Brown and Company (2000)

Hillary, Edmund *View from the Summit* Doubleday (1999)

Houston, Charles *Going Higher* Mountaineers (1998)

Hunt, John *The Ascent of Everest* Hodder and Stoughton (1953)

Hunt, John *Life is Meeting* Hodder and Stoughton (1978)

Izzard, Ralph *The Innocent on Everest* Hodder and Stoughton (1954)

Loomis, Chauncey *Weird and Tragic Shores* Alfred A. Knopf (1991)

Kane, Elisha *Arctic Explorations* Childs and Petersen (1856)

Kingsley, Mary *Travels in West Africa* MacMillan (1897)

Livingstone, D&C. *Expedition to the Zambezi* London (1865)

Lowe, George *Because it is There* Cassell (1959)

Messner, Reinhold *All 14 Eight-thousanders* Crowood (1999)

Messner, Reinhold *The Crystal Horizon* Crowood (1982)

McLintock, Leopold *A Narrative of the Discovery of the Fate of the Sir John Franklin and His Companions* John Murray (1860)

Morris, Jan *Coronation Everest* Faber and Faber (1958)

Nansen, Fridtjof *The First Crossing of Greenland* Longman (1890)

Park, Mungo *Travels in the Interior of Africa* London (1799)

Parry, William Edward *Journal of a Voyage for the Discovery of the North West Passage from the Atlantic to the Pacific in the Years 1819–20* John Murray (1821)

Peary, Josephine *My Arctic Journal* Longman 1893

Peary, Robert *The North Pole* Hodder and Stoughton (1910)

Peary, Robert *Secrets of Polar Travel* Century Co. (1917)

Rowell, Galen *In the Throne Room of the Mountain Gods* Allen and Unwin (1977)

Ruttledge, Hugh *Everest 1933* Hodder and Stoughton (1934)

Scott, Captain R. F. *Scott's last Expedition* Smith, Elder and Co. (1913)

Shackleton, Ernest *The Heart of the Antarctic* Heinemann (1909)

Shackleton, Ernest *South* Heinemann (1919)

Shipton, Eric *That Untravelled World* Hodder and Stoughton (1969)

Shipton, Eric *Blank on the Map* Hodder and Stoughton (1938)

Shipton, Eric *Upon that Mountain* Hodder and Stoughton (1943)

Shipton, Eric *Mount Everest Reconnaissance Expedition 1951* Hodder and Stoughton (1952)

Smith, Albert, *The Story of Mont Blanc* (1853)

Snow, Sebastian *Half a Dozen of the Other* Hodder and Stoughton (1972)

Stanley, H. M. *Through the Dark Continent* G. Newnes (1878)

Stanley, H. M. *How I Found Livingstone* Scribner (1872)

Stark, Freya *The Valley of the Assassins* Murray (1934)

Stark, Freya *The Southern Gates of Arabia* Murray (1936)

Stark, Freya *A Winter in Arabia* John Murray (1948)

Stuart, J. M. *Exploration in Australia* (1864)

Stuck, Hudson *The First Ascent of Denali* University of Nebraska (1914)

Thesiger, Wilfred *The Danakil Diary* HarperCollins (1996)

Thesiger, Wilfred *Desert, Marsh and Mountain* Collins (1979)

Thesiger, Wilfred *Arabian Sands* Penguin (1974)

Ullman, James R. *Man of Everest, The Autobiography of Tenzing* Transworld (1957)

Venables, Stephen *Alone at the Summit* Odyssey Books (1996)

Webster, Edward *Snow in the Kingdom* Mountain Imagery (2000)

Whymper, Edward *Scrambles Amongst the Alps* Murray (1871)
Whymper, Edward *In the Great Andes of the Equator* Murray (1911)
Whymper, Edward *A guide to Zermatt and the Matterhorn* Murray (1910)
Wilkins, Hubert *Under the North Pole* Benn (1931)
Worsley, F. A. *Shackleton's Boat Journey* Pimlico (1940)

Secondary Sources
Asher, Michael *Thesiger* Viking (1994)
Bryce, Robert M. *Cook & Peary, The Polar Controversy Resolved* Stackpole Books (1997)
Curran, Jim *K2, The History of the Savage Mountain* Coronet (1996)
Douglas, Ed *Tenzing, Hero of Everest* National Geographic (2003)
Dugard, Martin *Farther than Any Man* Allen and Unwin (2001)
Fiennes, Ranulph *Captain Scott* Coronet (2003)
Harper, Ken *Give Me Back my Father's Body* Blacklead Books (1986)
Herbert, Wally *The Noose of Laurels* Hodder and Stoughton (1989)
Huntford, Roland *Scott and Amundsen* Abacus (1979)
Huntford, Roland *Shackleton* Abacus (1985)
Hyde, Alexander *The Polaris Disaster or the Frozen Zone* Columbian Book Company (1874)
Jeal, Tim *Livingstone* Yale University Press (1973)
Kaufman, Andrew and Putnam, William *K2 and the 1939 Tragedy* The Mountaineers (1992)
Lovell, Mary S. *A Rage to Live* Little, Brown and Company (1998)
Lyall, Alan *The First Descent of the Matterhorn* Gomer (1997)
Malaurie, Jean *Ultima Thule* WM Norton (1990)
Murgatroyd, Sarah *The Dig Tree* Bloomsbury (2002)
Pakenham, Thomas *The Scramble for Africa* Abacus (1991)
Roberts, David *Annapurna, True Summit* Simon and Schuster (2000)
Shipman, Pat *To the Heart of the Nile* Bantam Press (2004)
Smythe, Frank *Edward Whymper* Hodder and Stoughton (1940)
Solomon, Susan *The Coldest March* Yale University Press (2001)
Unsworth, Walt *Everest, The Mountaineering History* Bâton Wicks (2000)
Wheeler, Sarah *Cherry* Jonathan Cape (2001)
Weems, John E. *Peary, The Explorer and the Man* Heinemann (1961)

Index